Lecture Notes in Computer Science 15540

The series Lecture Notes in Computer Science (LNCS), including its subseries Lecture Notes in Artificial Intelligence (LNAI) and Lecture Notes in Bioinformatics (LNBI), has established itself as a medium for the publication of new developments in computer science and information technology research, teaching, and education.

LNCS enjoys close cooperation with the computer science R & D community, the series counts many renowned academics among its volume editors and paper authors, and collaborates with prestigious societies. Its mission is to serve this international community by providing an invaluable service, mainly focused on the publication of conference and workshop proceedings and postproceedings. LNCS commenced publication in 1973.

Fouchal Hacène · Boumerdassi Selma ·
Renault Éric
Editors

Machine Learning for Networking

7th International Conference, MLN 2024
Reims, France, November 27–29, 2024
Revised Selected Papers

Springer

Editors
Fouchal Hacène
University of Reims Champagne-Ardenne
Reims, France

Boumerdassi Selma
CNAM
Paris, France

Renault Éric
ESIEE Paris - University Gustave Eiffel
Noisy-le-Grand, France

ISSN 0302-9743 ISSN 1611-3349 (electronic)
Lecture Notes in Computer Science
ISBN 978-3-032-00551-9 ISBN 978-3-032-00552-6 (eBook)
https://doi.org/10.1007/978-3-032-00552-6

This Springer imprint is published by the registered company Springer Nature Switzerland AG
The registered company address is: Gewerbestrasse 11, 6330 Cham, Switzerland

Preface

The rapid development of new network infrastructures and services has led to the generation of huge amounts of data, and machine learning now appears to be the best solution to process these data and make the right decisions for network management. The International Conference on Machine Learning for Networking (MLN) aims to provide a top forum for researchers and practitioners to present and discuss new trends in machine learning, deep learning, pattern recognition and optimization for network architectures and services. This year, MLN 2024 was hosted by Université de Reims Champagne-Ardenne, France. MLN 2024 was jointly organized by Lab-I* of Université de Reims Champagne Ardenne (URCA), the Laboratoire d'Informatique Gaspard-Monge (LIGM) and ESIEE Paris of Université Gustave Eiffel, and CNAM Paris.

The call for papers resulted in a total of 25 submissions from all around the world: Algeria, Bangladesh, France, India, Japan, Morocco and Singapore. All submissions were assigned to at least three members of the program committee for review. The program committee decided to accept 14 papers. Two invited papers were also presented.

The paper *Design and Evaluation of a Lightweight SDN Controller for Integrated Road and Rail Networks* by Dingyang LIU, Dereje Mechal Molla, Leo Mendiboure (Université Gustave Eiffel, France), Sassi MAALOUL (Université Gustave Eiffel, France and JUNIA Engineering School, France), Marion BERBINEAUS and Hakim BADIS (Université Gustave Eiffel, France) was awarded the prize for the best paper.

Four keynotes completed the program: *From Capturing Reality to Regenerating It: Transforming Autonomous Driving Simulation and Testing* by Gamal ELGHAZALY from Université du Luxembourg, Luxembourg, *Driving Energy Efficiency and Smarter User Behavior with IoT and Machine Learning* by Abbes AMIRA from University of Sharjah, UAE, *Transforming Urban Mobility and Vehicle Security: Exploring ICT and AI Advances in Smart City Networks* by Jalel BEN OTHMAN from Sorbonne Paris Nord University, and *On Reducing the Reliance on Data in Deep Learning-based Wireless Research* by Shiwen MAOS from Auburn University, USA.

We would like to thank all who contributed to the success of this conference, in particular the members of the Program Committee and the reviewers for carefully reviewing the contributions and selecting a high-quality program. Our special thanks go to the members of the Organizing Committee for their great help.

We hope that all participants enjoyed this successful conference.

Cyril Rabat
Renault Éric

Organization

General Chairs

Cyril Rabat	Lab-I*, URCA, France
Éric Renault	ESIEE Paris – Université Gustave Eiffel, France

Technical Program Chairs

Selma Boumerdassi	CNAM, France
Hacène Fouchal	Lab-I*, URCA, France

Technical Program Committee

Antonio Cianfrani	Sapienza University of Rome, Italy
Anwer Al-Dulaimi	EXFO, Canada
Cherkaoui Leghris	Hassan II University, Morocco
Claudio A. Ardagna	Università degli Studi di Milano, Italy
Eman-Yaser Daraghmi	PTUK, Palestine
Eun-Sung Jung	Hongik University, South Korea
Florent Nolot	Lab-I*, Université de Reims Champagne-Ardenne, France
Fred Aklamanu	Nokia Bell Labs, France
Gianfranco Nencioni	University of Stavanger, Norway
Gloria Elena Jaramillo	DFKI, Germany
Hamid Anvari	University of Alberta, Canada
Hella Kaffel Ben Ayed	Université de Tunis El Manar, Tunisia
Jean-Charles Grégoire	INRS, Canada
Jean-François Bercher	ESIEE Paris, France
Luiz Bittencourt	University of Campinas, Brazil
Mai Trang Nguyen	Sorbonne Paris Nord University, France
Mariam Kiran	Lawrence Berkeley National Laboratory, USA
Martine Wahl	Université Gustave Eiffel, France
Marwane Ayaida	UPHF, France
Maxim Bakaev	NSTU, Russia
Miki Yamamoto	Kansai University, Japan
Mohamed Belaoued	University of Constantine 1, Algeria

Mouaasoui Boubakeur	Université de Msila, Algeria
Nadjib Achir	Université Sorbonne Paris Nord, France
Naïla Bouchemal	ECE, France
Nadir Mahammed	École Supérieure en Informatique de Sidi Bel Abbès, Algeria
Nicola Bena	Università degli Studi di Milano, Italy
Nirbhay Chaubey	Ganpat University, India
Olaf Maennel	Tallinn University of Technology, Estonia
Ouadoudi Zytoune	Université Ibn Tofail, Morocco
Patrick Sondi	IMT Nord Europe, France
Paulo Pinto	Universidade Nova de Lisboa, Portugal
Ruben Milocco	Universidad Nacional del Comahue, Argentina
Sabine Randriamasy	Nokia Bell Labs, France
Smain Femmam	Université de Haute Alsace, France
Stavros Shiaeles	University of Portsmouth, UK
Toufik Ahmed	INP Bordeaux, France
Van Khang Nguyen	Hue University, Vietnam
Van Long Tran	Enlab Software, Vietnam
Viet Hai Ha	Hue University, Vietnam
Vinod Kumar Verma	SLIET, India
Yousef-Awwad Daraghmi	PTUK, Palestine

Publicity Chair

Ramzi Boutahala	URCA, France

Sponsoring Institutions

CNAM, France
ESIEE Paris, France
Université Gustave Eiffel, France
Université de Reims Champagne-Ardenne, France

Steering Committee

Selma Boumerdassi	CNAM, France
Paul Mühlethaler	Inria, France
Éric Renault	ESIEE Paris – Université Gustave Eiffel, France

Contents

Learning Per-Flow SD-WAN Load-Balancing Policies 1
Hocine Abdelhamid, David Jia, and Vania Conan

Survey on Federated Learning in Smart Healthcare 14
Assia Hammamouche, Djamila Zamouche, Lyes Khoukhi, and Said Hamamouche

Complex Communication Networks Management with Distributed AI: Challenges and Open Issues 32
Christina Alhachem, Mounir Kellil, and Abdelmadjid Bouabdallah

A Framework for Global Trust and Reputation Management in 6G Networks: Position Paper 55
Bac Trinh-Nguyen, Sara Berri, Sin G. Teo, Tram Truong-Huu, and Arsenia Chorti

DRL Framework for Minimizing Beam Switching Time and Maintaining QoS in 6G-V2X Base Stations 73
Ali Belgacem and Abbas Bradai

Reducing BLE Energy Loss in Busy 2.4GHz Band 85
Morgane Joly, Éric Renault, and Fabian Rivière

Leveraging SHAP to Advance the Robustness of Large Language Models 97
Mohamed Yacine Djema, Hacene Fouchal, and Olivier Flauzac

Keyword-Driven Email Classification: Leveraging Machine Learning Techniques 102
Jiahui Xiang, Osman Salem, and Ahmed Mehaoua

Predicting Intents: ARMA-Based Modeling 116
Nagham Hachem and Éric Renault

Design and Evaluation of a Lightweight SDN Controller for Integrated Road and Rail Networks 133
Dingyang Liu, Dereje Mechal Molla, Leo Mendiboure, Sassi Maaloul, Marion Berbineau, and Hakim Badis

PiPS: An Effective Strategy and Approach for Privacy in Public Surveillance 149
Sajid Ahmed and Noriaki Yoshiura

A Comprehensive Review of Deep Learning Approaches for Tomato Leaf Diseases Detection and Classification in Smart Agriculture 170
Oussama Nabil and Cherkaoui Leghris

A Review on Advancement in PEM Fuel Cell Diagnosis Based on Machine Learning Techniques 180
Hichem Kahia, Selma Boumerdassi, Aissa Belmeguenai, Abderrahmane Herbadji, and Djamel Herbadji

GPS Spoofing Attack Against UAVs: A Timeseries Dataset Case Study 195
Mouzai Mustapha and Riahla Mohamed Amine

Author Index 209

Learning Per-Flow SD-WAN Load-Balancing Policies

Hocine Abdelhamid, David Jia(✉), and Vania Conan

Thales SIX, Gennevilliers, France
{Hocine.Abdelhamid,David.Jia,Vania.Conan}@thalesgroup.com

Abstract. Software Defined Wide Area Network (SD-WAN) offers an overlay solution for businesses to interconnect several remote branches to a common enterprise cloud. The SD-WAN overlay is built over several WAN transport connections (e.g., MPLS, 4G/5G or Internet services). The network operator then defines traffic engineering policies to balance the traffic load over the different WAN underlays. In this work we consider streams of IP packets, as in open-flow, which we call flows. We explore per-flow load-balancing which offers a more versatile and fine grained policy definition to the network operator. We formulate the load-balancing problem in a control theoretic framework. We compare two approaches to solve the problem: an optimal closed-form solution to the Model Predictive Control (MPC) formulation and a Deep Reinforcement Learning (DRL) Proximal Policy Optimization (PPO) based algorithm. We show that the approximate DRL algorithm reaches the MPC solution with less than 5 ‰ error on the load balancing ratio for the Poisson traffic case we considered and that it can address a large class of mixed traffic through Machine Learning generalisation.

Keywords: Software Defined Wide Area Network (SD-WAN) · Deep Reinforcement Learning · Enterprise Network · Per-Flow Load Balancing

1 Introduction

Businesses increasingly rely on cloud-based operations which requires all branches to be efficiently connected to one (or several) clouds through Wide Area Network (WANs). The WAN overlay networks are traditionally built by leasing complex and costly Multiprotocol Label Switching (MPLS) services from telecommunications operators. Software Defined Wide Area Network (SD-WAN) is rapidly being adopted as a transport agnostic overlay solution [14], with the global SD-WAN market predicted to grow at an annual growth rate of 31.9% from 2022 to 2027 [8]. Hybrid SD-WAN solutions, mixing MPLS services with xDSL, 4G/5G or cable Broadband Internet services reduce costs or improve resilience. Its centralised control simplifies the configuration and management of the network devices. The behaviour of the WAN is thus defined by networking

F. Hacène et al. (Eds.): MLN 2024, LNCS 15540, pp. 1–13, 2026.
https://doi.org/10.1007/978-3-032-00552-6_1

policies customised by the end user businesses. Gartner considers that, by 2026, 70% of enterprises will have implemented SD-WAN networks, compared with approximately 45% in 2021 [5].

Current SD-WAN products for enterprise networks offer network management solutions that allow enterprises to tailor it to their business intents and traffic demands. These policies are implemented as Traffic Engineering rules that govern the routing of the enterprise traffic. SD-WAN rules identify traffic of interest and select the outgoing interface based on a strategy and the condition of the route or link between two devices. Many performance metrics are used to evaluate the link conditions and many strategies are available: e.g. manual or automatic, best quality or lowest cost, traffic shaping or DSCP-tag based [3].

In practice this leads to static Traffic Engineering (TE) rules that apply to aggregated IP traffic, or IP traffic Quality of Service (QoS) classes. On the other hand, Software Defined Networking (SDN) technologies such as OpenFlow are able to process and forward traffic from OSI layers 2 or 3 up to application layers (layer 4 and above). This paves the way for a more fine-grained steering of traffic in the network devices: each application layer session can thus be forwarded through any of the available outgoing SD-WAN connections [9]. In this approach the SD-WAN forwarding rules map each session to an outgoing network interface.

In this work we explore the benefits of such per-flow SD-WAN policies. The policies are defined as to optimise business criteria defined by the network operator at application level. The policies are computed automatically and enforced at the edge devices in the enterprise network.

The main contributions of this work are:

- We express SD-WAN policies at flow/session level and formulate the problem of finding an optimal policy in a control theoretic framework.
- We provide a closed-form solution to the load-balancing problem in the case of simple Poisson traffic.
- We introduce a Deep Reinforcement Learning (DRL) solution to learn the load-balancing policy from observed data traffic.
- We show that the DRL algorithm learns near optimal policies in the case of simple traffic and is able to generalise the policies to cope with more complex mixed traffic.

The rest of this paper is organized as follows. Section 2 discusses the general background and related literature. Section 3 formulates per-flow SD-WAN as control problem. It provides a closed-form solution to load-balancing policy computation and introduces a DRL algorithmic solution. Section 5 focuses on evaluating the quality of the DRL solution against the closed-form optimal one. Furthermore it demonstrates the generalisation property of the DRL solution. Section 6 discusses the main achievements and concludes the paper.

2 Related Work

A first body of work addresses SD-WAN policies as Traffic Engineering rules. In this context, the SD-WAN overlay network can be designed so as to improve its robustness or availability. Fault tolerance of the overlay is addressed in [15] which uses an Integer Linear Program (ILP) formulation to improve scalability and management of the overlay by a careful monitoring of the underlay links and global optimisation of the routes. In [12] DRL is used in an SD-WAN context to redirect the traffic to maximize availability. They study different DRL algorithms with different reward definitions and compare the to baseline TE algorithms. In this work we propose a different formulation of the SD-WAN traffic distribution problem, based on control-theory. We make the link with a DRL resolution of the problem.

Since SD-WAN services operate based on application-level policies, another important body of work focuses on finer grained flow or session level mechanisms. The major objective is to maximise the connection utilisation over a single (or aggregated) link. The earliest works address inter-data center data traffic and target Best Effort traffic only and the solutions consist in scheduling the data exchanges round the clock [6,7]. Providing higher levels of application visibility also leads to the possibility of incorporating finer grained policies which apply to individual flows. [2] addresses an SD-WAN technique of scheduling mix-flows with different QoS requirements. They consider a single link but a mix of deadline and non-deadline flows and propose an information-agnostic mix-flows scheduling mechanism to minimize flow completion time. In this work we focus on load-balancing policies, with the aim to improve the average traffic imbalance across the different SD-WAN links.

3 Problem Statement

3.1 System Model

To address per-flow load-balancing, we focus on the SD-WAN module of an edge device. The device is connected to two WAN links L_1 and L_2. For each outgoing flow, a controller decides to which outgoing interface the flow packets should be forwarded to. Once the first packet of a flow is sent on an interface, all following packets are sent to the same interface (to ensure as much as possible the consistent ordering of the delivered packets). The load-balancing policy implemented by the controller aims to keep a certain proportion ρ of (total) outgoing traffic on L_1 (and thus $1-\rho$ on L_2). With $\rho = 0.5$ traffic is distributed evenly on the two interfaces. If the two links have different costs or QoS properties, one can wish to favor one link over the other (the cheaper or the better quality one) in which case one would choose to distribute traffic unevenly (e.g., with $\rho = 0.75$).

The SD-WAN module is also monitoring the outgoing load on each link every ΔT time interval. This may be performed by an SDN module which counts the outgoing packets, and reports every ΔT the corresponding aggregated load to the SD-WAN module (i.e., x_t for L_1 and y_t for L_2) [9,13]. The objective of the controller is to keep the traffic on L_1 as close as possible to ρ and the traffic on L_2 as close as possible to $1-\rho$.

Its performance can be measured by quadratic criterion J:

$$J = (\rho_t^1 - \rho)^2 + (\rho_t^2 - (1-\rho))^2, \tag{1}$$

where $\rho_t^1 := \frac{x_t}{x_t+y_t}$ and $\rho_t^2 := 1 - \rho_t^1 = \frac{y_t}{x_t+y_t}$ are the aggregated load ratios of links L_1 and L_2 at t, respectively.

3.2 Model Predictive Control

We cast the above load-balancing problem into a probabilistic control theory framework: at time interval t and for the following ΔT time interval, the SD-WAN module decides to send the new flow to link L_1 with probability u_t ($0 < u_t < 1$). A randomized round robin policy corresponds to a constant value $u_t = \rho$. But this creates some load balancing imbalance in the network loads since the sessions have different service times [1]. This is confirmed by the load dependence of the optimal solution in (3).

We thus introduce a Model Predictive Control (MPC) approach to compute the optimal value of u_t at each time t (incremented every ΔT) knowing the load on each outgoing interface (x_t and y_t) in the previous time interval (see Fig. 1).

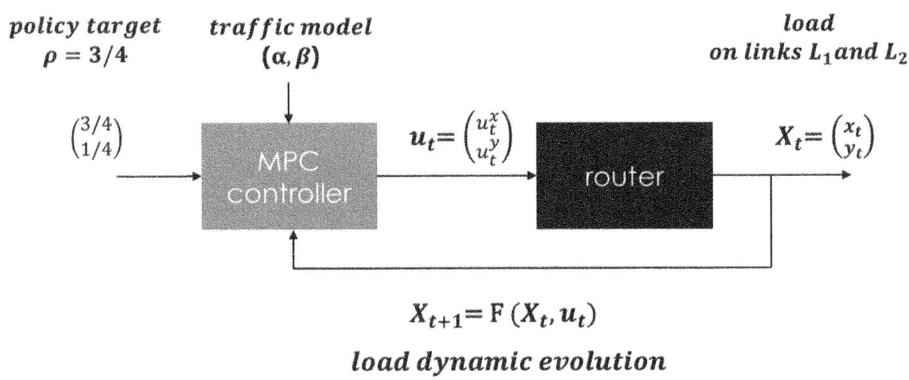

Fig. 1. Control loop for the load balancing model.

To that end, we consider sessions are generated by an $M/M/1$ queue with two parameters λ and μ. λ is the rate of incoming sessions (Poisson arrivals) and μ characterizes their average duration (exponential service time). The Poisson distribution has the advantage to be simple to set up and manipulate mathematically. It is a coarse first hypothesis to represent Internet traffic dynamics at the session level [4]. The study of more complex models is left for future work. More importantly, it yields a mathematical closed-form solution for the load-balancing policy which provides a baseline solution against which we may compare our trained RL agent.

Applying the above load-balancing policy with probability u_t leads to the following evolution equations for the traffic loads:

$$\mathbf{X_{t+1}} = F(\mathbf{X_t}, \mathbf{U_t}) = \begin{pmatrix} x_{t+1} = \alpha x_t + \beta u_t \\ y_{t+1} = \alpha y_t + \beta(1 - u_t) \end{pmatrix} \tag{2}$$

with
$\alpha = 1 - \mu\Delta T$: persistence rate of ongoing flows
$\beta = \lambda\Delta T$: arrival rate of new flows

Minimising (1) leads to the predictive law:

$$u_t = \frac{1}{2}\left(1 + \frac{\alpha}{\beta}(y_t - x_t) + \frac{2\rho - 1}{\beta}(\alpha(x_t + y_t) + \beta)\right) \tag{3}$$

This provides a closed-form solution for updating command u_t at every ΔT once the traffic observation is done. Note however that this requires knowledge of the incoming traffic distribution since formula (3) depends on λ and μ. To circumvent this limitation we next turn to machine learning.

4 Proposed Solution

4.1 Reinforcement Learning Formulation

Reinforcement Learning (RL) is a machine learning approach where an agent learns to make decisions by interacting with its environment. The RL problem is modeled as a finite Markov Decision Process (MDP), defined by three variables provided at each time step t: states, actions, and rewards. The state describes the current condition of the environment. Based on this state, the agent takes an action that influences the environment. The environment provides feedback as rewards, numerical values the agent aims to maximize over time. [11].

In the context of per-flow load-balancing, RL is particularly suitable for fluctuating environments. The agent can dynamically adjust every step t, whose duration is ΔT time interval, the distribution of the flow packets into the WAN links using measures from the SD-WAN module.

States. The state space consists of the discrete number of aggregated loads x_t,y_t in the links L_1, L_2, respectively. At each step t of duration ΔT, new state values x_t,y_t are sampled.

Action. The action space u_t^{RL} represents the probability of the load-balancing strategy for the outgoing traffic on L_1 at each step t of duration ΔT. Its aim is to guarantee an average load proportion of traffic in L_1 of ρ.

Reward. The load-balancing goal is to have links L_1 and L_2's load close to ρ and $1-\rho$, respectively. To this end, the reward is calculated at each time step t as the difference between the load ratio ρ_t^1 and link L_1's load objective ρ. Moreover, we penalize the reward at t if it is lower than the reward calculated at $t-1$ to ensure a monotonic improvement. Formally, the reward function can thus be written as:

$$r_t = \begin{cases} \dfrac{1}{|\rho_t^1 - \rho|} & \text{if } r_t > r_{t-1} \\ K\left(|\rho_t^1 - \rho|\right) & \text{else.} \end{cases} \tag{4}$$

where K is the penalty coefficient and is set as -1000 in the numerical simulations.

4.2 Proposed Deep Reinforcement Learning Architecture

Since the previously described MDP has a high-dimensional state space, we use a DRL approach to approximate the optimal policy. To this end, we choose to employ the Proximal Policy Optimization (PPO) algorithm [10] due to its suitability for continuous action spaces, unlike conventional approaches such as Deep Q-network (DQN) [11].

PPO is an online policy gradient method in DRL, designed to improve training stability and performance. It works by ensuring the policy update is within a specified range to stabilize learning. PPO achieves this through a surrogate objective function that penalizes large deviations from the current policy, enabling more reliable and efficient learning, especially in fluctuating environments like our per-flow SD-WAN load-balancing context.

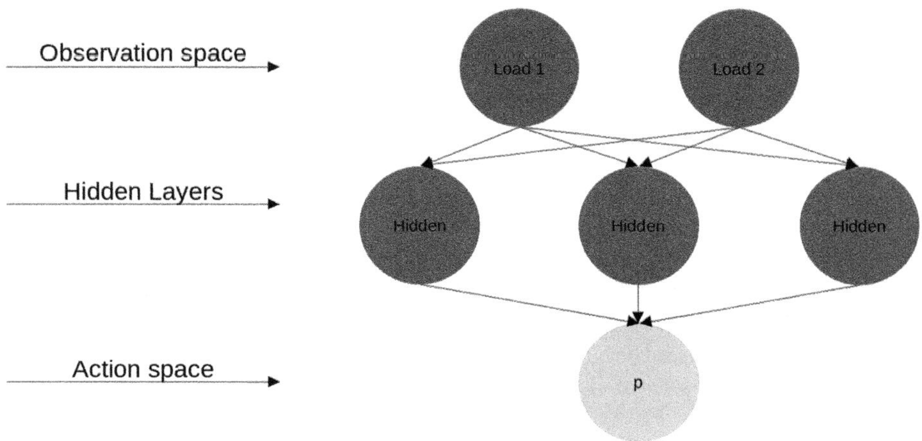

Fig. 2. Neural network architecture used for the load-balancing task when using PPO, the MDP is composed of two discrete values of states and one continue value of action.

To learn this surrogate function, we use a neural network which is illustrated in Fig. 2. Its architecture consists of an observation space, hidden layers, and an action space. The output is a single node representing the load-balancing ratio ρ of distributing traffic to one of the links. At each neuron, the activation function is a ReLU function, except at the output layer where it is a sigmoid function to obtain a value between 0 and 1.

4.3 Training Approach for Generalization

The works described in the previous section focused on a DRL training strategy with fixed Poisson process parameters λ and μ. We evaluated this model on sessions with identical parameters λ and μ. However, real-world applications require a more flexible model able to handle various traffic sessions without prior knowledge of their specific parameters, i.e. λ and μ.

Therefore, to enhance the DRL model's generalization capabilities for various λ and μ values, we introduce a novel training approach involving alternating between different λ and μ configurations. Our extensive simulations demonstrate that this strategy, specifically alternating every 1000 iterations rather than every iteration, yields superior training results. Indeed, this approach allows the model to effectively learn from each configuration, leading to improved generalization. In this paper, we refer to DRL model with fixed parameters and the novel one as conventional DRL and generalized DRL, respectively.

5 Numerical Results

5.1 Simulation Environment and RL Performance

In this section, we compare the performance of the MPC approach, and our proposed DRL models in terms of mean and standard deviation (std) loads ratio allocated in link L_1.

We opt for an event-based simulation which allows us to train our agent faster than if it was real-time traffic. An event is defined as either incoming or outgoing sessions, which follow a Poisson process of parameters λ and μ, to the SD-WAN controller. We assume that the SD-WAN module monitors the outgoing load on each link every $\Delta T = 1s$. The RL environment and the PPO have been implemented using Gym and Stable Baselines3 Python library, respectively. The PPO hyperparameters were chosen with their default parameters. The simulations were conducted on a standard CPU machine equipped with an Intel Core i7-11850H processor, running at 2.5 GHz with 16 GB of RAM.

Figure 3 illustrates the cumulative reward of the DRL model over training iterations. The upward trend indicates effective DRL training.

5.2 MPC vs DRL

In this section, we compare the allocated flows using the DRL's agent policy against the MPC's one at each step t. The parameters are set as $\lambda = 20, \mu = 0.2$ and the DRL agent is trained on sessions with these same parameters.

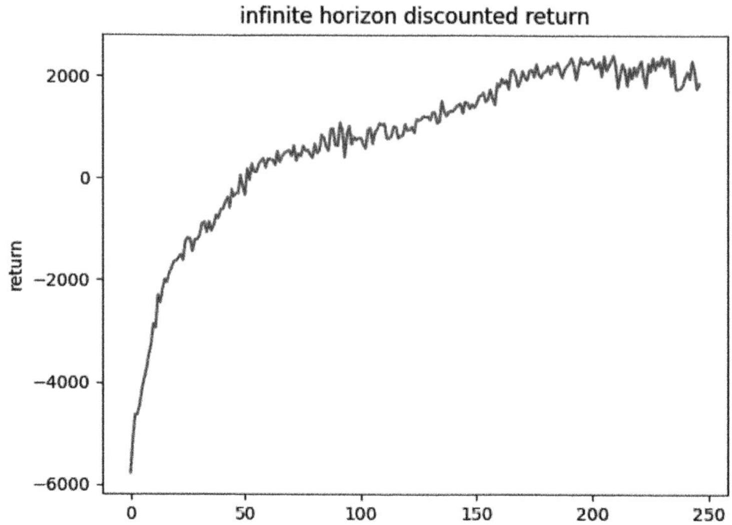

Fig. 3. Cumulative reward of the DRL model, versus epochs.

Figure 4 shows the load ratio in link L_1 using the DRL and the MPC policy, versus the events for 40000 events, i.e. 20000 sessions. It also shows the calculated mean and std ratio over all the events. From this figure, we can observe that the MPC and the DRL's ratio both oscillate around the expected proportion $\rho = 0.75$ and that their mean ratio is close to ρ. Moreover, the std of the DRL agent is close to the MPC one. These observations show that the trained DRL model has similar performance to the closed-form solution provided by the MPC.

5.3 Generalisation and Mixed Traffic

In this section, we study the proposed DRL solutions' generalization for different kind of traffic to reach the load ratio $\rho = 0.75$. For clarity, our simulations parameterize $1/\mu$ instead of μ, as $1/\mu$ represents the more intuitive value of average session duration. As mentioned in Sect. 4.3, the first DRL model is trained on Poisson traffic for $\lambda = 20, 1/\mu = 5$ while the second model alternates its training between $\lambda = 20$ and $\lambda = 80$, and between $1/\mu = 1.25$ and $1/\mu = 5$ every 1000 iterations. To evaluate them, we compare our trained load-balancing models to the MPC one by generating Poisson arrivals sessions with varying parameters λ and μ. We then study their performance in terms of the mean and std aggregated load ratios over 10^6 Poisson sessions.

Figures 5a and 5b show the mean and the std of the aggregated load ratios, respectively, in L_1 of several policies for multiple values of λ. In Fig. 5a, we observe that the MPC mean load ratio is steady around 0.75 for varying λ and has better performance than the DRL models, which was expected since the MPC provides the theoretical optimal solution given the parameter λ. Furthermore, we can see that both DRL solutions have decreasing values of load ratio

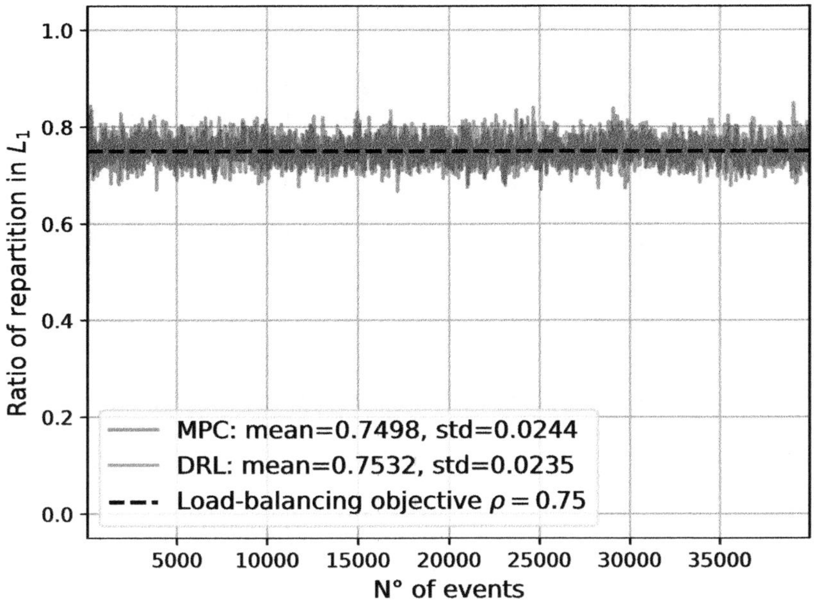

Fig. 4. Aggregated load ratios of link L_1, versus the number of events for $\lambda = 20, 1/\mu = 5$ and $\Delta T = 1$s.

between 0.74 and 0.76 for increasing λ. The conventional DRL model shows performance comparable to the optimal solution near $\lambda = 20$, but demonstrates lower performance for other λ values, as anticipated. We can observe for $\lambda > 30$ that the proposed generalized DRL consistently outperforms the regular DRL model since the mean load ratio is closer to 0.75. This suggests that training the DRL model on sessions with $\lambda = 20$ and $\lambda = 80$ is beneficial for a wider range of λ. For λ between 5 and 20, the conventional DRL approach shows better results since its training seems to have focused more on lower values of λ. In Fig. 5b, we can see that the std of all the approaches decrease for increasing λ values. For $\lambda > 40$, the std of the generalized DRL is close to the MPC's and outperforms the conventional DRL's.

Figures 6a and 6b show the mean and the std of the aggregated load ratios, respectively, in L_1 of several policies for multiple values of $1/\mu$. In Fig. 6a, we can see that the MPC mean load ratio is steady around $\rho = 0.75$, as expected. Both DRL models show decreasing values of mean load ratio for increasing $1/\mu$. Their performance is suboptimal for $1/\mu < 1$. This can be attributed to the short session duration: the SD-WAN module, which reports outgoing packet counts every ΔT, may miss sessions that are too brief. In addition, the conventional DRL model shows satisfactory performance near $1/\mu = 5$, as expected. The generalized DRL model demonstrates for $1/\mu > 8$ better performance, highlighting the effectiveness of our generalization training.

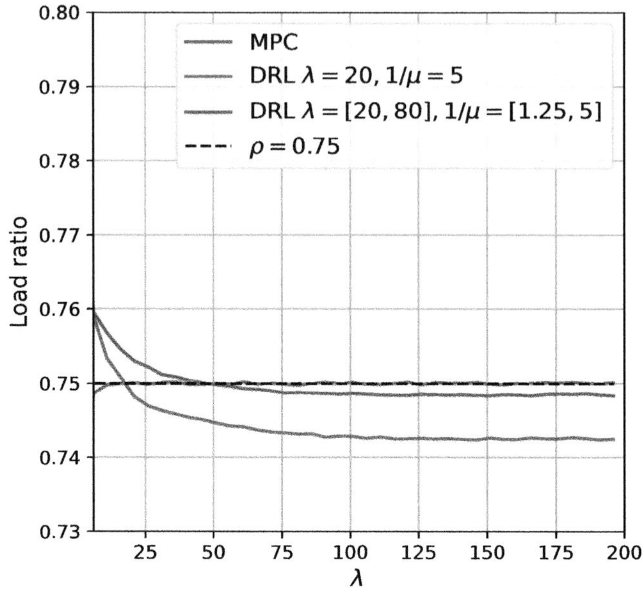

(a) Mean aggregated load ratio, versus λ

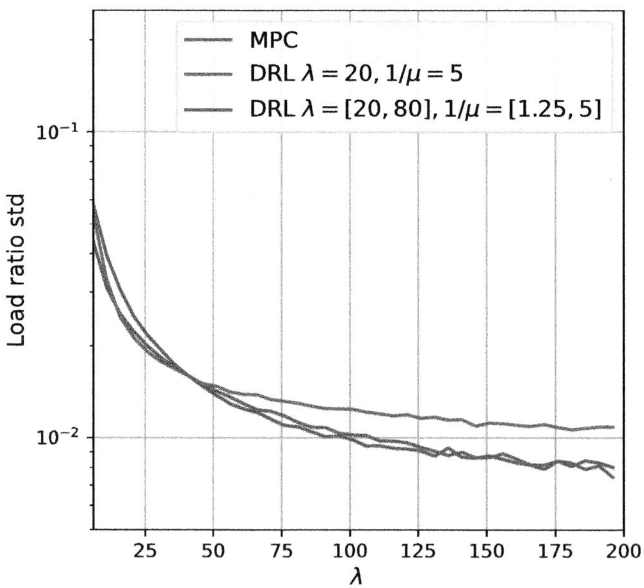

(b) Std aggregated load ratio, versus λ

Fig. 5. Mean and std aggregated load ratio, versus λ, for $1/\mu = 5$.

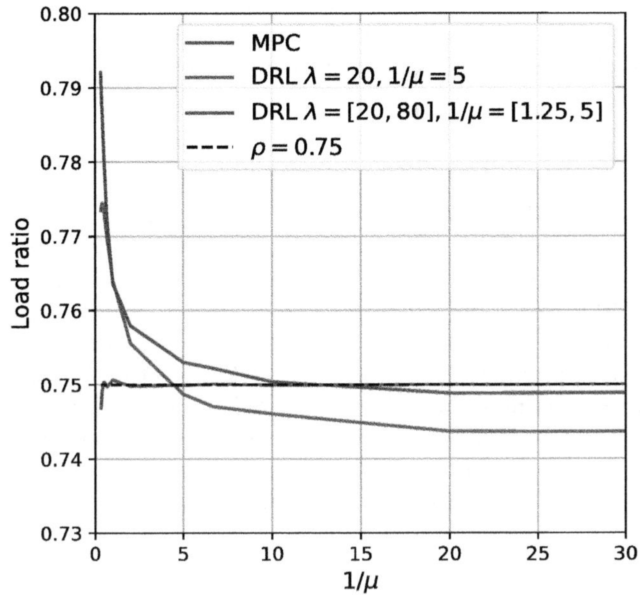

(a) Mean aggregated load ratio, versus $1/\mu$

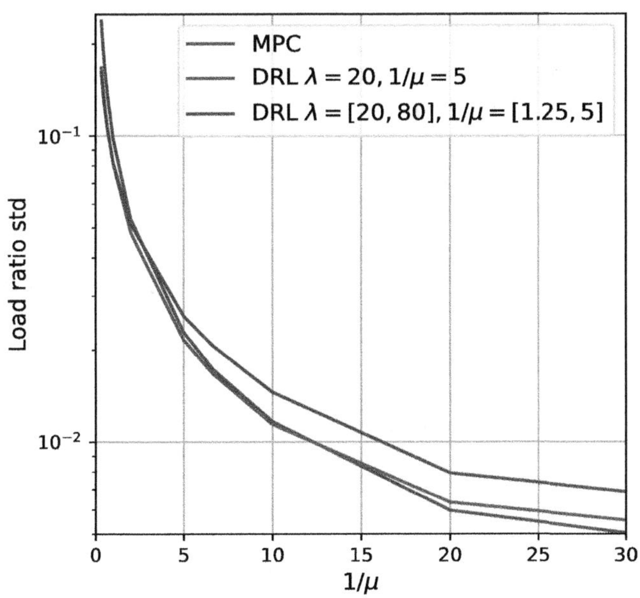

(b) Std aggregated load ratio, versus $1/\mu$

Fig. 6. Mean and std aggregated load ratio, versus $1/\mu$, for $\lambda = 20$.

In Fig. 6b, the generalized DRL model shows for $1/\mu > 8$: i) lower std compared to the conventional one, and ii) comparable std to the MPC. These findings reinforce the robustness and stability of the proposed generalized DRL approach.

6 Conclusion

In this work we have addressed per-flow load-balancing in an SD-WAN context. We cast the problem in a control theoretic framework. We demonstrated the benefit of using a PPO based Machine Learning technique to solve the problem: it generalises well to mixed traffic and reaches near optimal solutions when compared to closed-form optimal solution in the case of simple traffic. This is a promising first step which shows that the DRL algorithm is able to learn more complex traffic dynamics than the simple MM1 model. Future work includes considering how such an approach would perform with bursty traffic. One can also explore how more elaborate QoS based policies, including ones addressing latency, would also be expressed in such a framework.

References

1. Barbette, T., Wu, E., Kostić, D., Maguire, G.Q., Papadimitratos, P., Chiesa, M.: Cheetah: a high-speed programmable load-balancer framework with guaranteed per-connection-consistency. IEEE/ACM Trans. Netw. **30**(1), 354–367 (2022). https://doi.org/10.1109/TNET.2021.3113370
2. Dong, X., Wei, N., Zhou, X., Zhao, L., Li, K.: QIMS: QoE-centric information-agnostic mix-flows scheduling in SD-WAN. In: 2019 IEEE 21st International Conference on High Performance Computing and Communications; IEEE 17th International Conference on Smart City; IEEE 5th International Conference on Data Science and Systems (HPCC/SmartCity/DSS), pp. 1274–1281 (2019). https://doi.org/10.1109/HPCC/SmartCity/DSS.2019.00178
3. Fortinet: Fortios 7.2.5 administration guide. Tech. rep., Fortinet (2023)
4. Frost, V., Melamed, B.: Traffic modeling for telecommunications networks. IEEE Commun. Mag. **32**(3), 70–81 (1994). https://doi.org/10.1109/35.267444
5. Gartner: Magic quadrant for network services, global. Tech. rep., Gartner. Accessed 22 Feb 2023
6. Jain, S., et al.: B4: experience with a globally-deployed software defined wan, vol. 43, pp. 3–14 (2013). https://doi.org/10.1145/2486001.2486019
7. Kandula, S., Menache, I., Schwartz, R., Babbula, S.: Calendaring for wide area networks. ACM SIGCOMM Comput. Commun. Rev. **44** (2014). https://doi.org/10.1145/2619239.2626336
8. Markets, M.: Software-defined wide area network (SD-WAN) market by component (solutions and services), deployment mode (on-premises and cloud), organization size, end user (service providers and verticals) and region - global forecast to 2027. Tech. rep., Markets & markets (2023)
9. Neghabi, A.A., Jafari Navimipour, N., Hosseinzadeh, M., Rezaee, A.: Load balancing mechanisms in the software defined networks: a systematic and comprehensive review of the literature. IEEE Access **6**, 14159–14178 (2018). https://doi.org/10.1109/ACCESS.2018.2805842

10. Schulman, J., Wolski, F., Dhariwal, P., Radford, A., Klimov, O.: Proximal policy optimization algorithms (2017)
11. Sutton, R.S., Barto, A.G.: Reinforcement Learning: An Introduction, second edn. The MIT Press (2018). http://incompleteideas.net/book/the-book-2nd.html
12. Troia, S., Sapienza, F., Varé, L., Maier, G.: On deep reinforcement learning for traffic engineering in SD-WAN. IEEE J. Sel. Areas Commun. **39**(7), 2198–2212 (2021). https://doi.org/10.1109/JSAC.2020.3041385
13. Troia, S., Zorello, L.M.M., Maralit, A.J., Maier, G.: SD-WAN: an open-source implementation for enterprise networking services. In: 2020 22nd International Conference on Transparent Optical Networks (ICTON), pp. 1–4 (2020). https://doi.org/10.1109/ICTON51198.2020.9203058
14. Yang, Z., Cui, Y., Li, B., Liu, Y., Xu, Y.: Software-defined wide area network (SD-WAN): architecture, advances and opportunities. In: 2019 28th International Conference on Computer Communication and Networks (ICCCN), pp. 1–9 (2019). https://doi.org/10.1109/ICCCN.2019.8847124
15. Zad Tootaghaj, D., Ahmed, F., Sharma, P., Yannakakis, M.: Homa: an efficient topology and route management approach in SD-WAN overlays. In: IEEE INFOCOM 2020 - IEEE Conference on Computer Communications, pp. 2351–2360 (2020). https://doi.org/10.1109/INFOCOM41043.2020.9155503

Survey on Federated Learning in Smart Healthcare

Assia Hammamouche[1,2(✉)], Djamila Zamouche[2], Lyes Khoukhi[1], and Said Hamamouche[3]

[1] GREYC Laboratory, Normandy University, 14000 Caen, France
assiahamamouche@gmail.com
[2] LIMED Laboratory, Bejaia University, 06000 Bejaia, Algeria
[3] Paris Cité University, 75013 Paris, France

Abstract. Federated Learning (FL) in smart healthcare systems is an innovative approach to distributed analysis of medical data while preserving confidentiality. By enabling collaborative learning without compromising patient privacy, FL offers a promising solution to the growing need for more efficient and secure medical applications. Researchers have been investigating various FL approaches to address the challenges in smart healthcare, with a focus on improving data use while maintaining high privacy assurance levels. In this paper, an analysis of aggregation models and secure aggregation techniques in smart healthcare to ensure privacy is presented. Then, we provide a comprehensive review of recent advances and applications of FL in smart healthcare systems. We analyze several FL-based approaches proposed in the literature, highlighting their strengths, limitations and potential impact on healthcare systems performance. In addition, we propose a taxonomy to classify the existing FL-based approaches.

Keywords: Smart Healthcare Systems · Federated Learning · Distributed Analysis · Aggregation · Data Privacy

1 Introduction

The evolution of digital technologies is deeply transforming the healthcare sector and giving rise to innovative smart healthcare systems. By integrating the Internet of Things (IoT), Artificial Intelligence (AI), and massive data analysis, these systems aim to revolutionize healthcare services [1]. Exploiting these technologies, smart healthcare systems concretely improve the quality of care through real-time monitoring and AI-assisted medical imaging. In addition, they enable personalized treatment based on the analysis of real-time data, and optimize the management of medical resources via decision support systems, which leads to more efficient medical practice. However, despite this progress, smart healthcare systems are facing major challenges that are preventing their full potential. Among these challenges, they include the increasing confidentiality concerns related to medical data, the lack of comprehensive datasets, and the high

F. Hacène et al. (Eds.): MLN 2024, LNCS 15540, pp. 14–31, 2026.
https://doi.org/10.1007/978-3-032-00552-6_2

costs associated with training AI models on these data. Issues such as security threats, including data integrity attacks, are also significant concerns that must be addressed to ensure reliable smart healthcare solutions [2]. These issues lead to crucial considerations about the long-term effectiveness of current approaches of smart healthcare. In this context, Federated Learning (FL) is emerging as a promising solution to address the challenges posed by smart healthcare systems, especially in terms of data confidentiality and access to diverse datasets [3]. Indeed, this innovative approach enables machine learning models to be trained on decentralized data, without requiring them to be transferred to a central server. Thus, FL improves privacy protection, makes access to diversified data easy, and reduces the costs associated with data centralization and processing. Recent studies have further explored how integrating blockchain into FL can enhance data integrity and trustworthiness, which is critical for secure healthcare systems [4]. Moreover, FL enables healthcare institutions to collaborate on large-scale research projects where knowledge can be shared without compromising patient confidentiality [5] [6].

There are previous efforts that have already been focused on the use of FL in the field of smart healthcare. As a result, multiple approaches have been proposed to make the most of FL, demonstrating satisfactory performance in various healthcare applications. For instance, the FL has been successfully applied in several studies to develop predictive models for disease progression, drug discovery and personalized treatment plans, often achieving accuracy that is comparable to that of centralized learning methods while preserving data confidentiality [7]. In the field of medical imaging analysis, FL has demonstrated significant capabilities in improving diagnostic efficiency in diseases such as breast cancer and retinal diseases [8]. Additionally, research has proposed innovative detection techniques using machine learning models to enhance security in healthcare and other fields, such as advanced attack detection frameworks in Software-Defined Networks (SDN) [9]. Furthermore, efficient data management and offloading approaches are being developed to optimize mobile cloud computing resources, which may benefit the healthcare sector by improving data handling and reducing latency in mobile healthcare applications [10]. The performance of these FL models often match or even surpass traditional centralized approaches in terms of accuracy, while significantly reducing privacy risks and data transfer costs.

Some surveys on the integration of FL in healthcare systems [11–13] have been carried out. Our study in the framework of this paper stands out from other reviews on FL by providing an analysis of aggregation models and Secure Aggregation (SA) techniques in smart healthcare to ensure privacy. Additionally, we propose an innovative taxonomy that classifies FL applications into four categories: medical image processing, managing electronic health records (EHRs), drug discovery and development, and patient monitoring and remote healthcare. An in-depth study including the most relevant and recent solutions in the field of FL applied to smart healthcare systems is provided. We analyze in detail the effectiveness of each approach, highlighting their strengths, limitations and applicability in various healthcare contexts.

The remaining of this paper is organized as follows. We give the key motivations of the use of FL in smart healthcare in Sect. 2. Section 3 provides the key principles underlying data partitioning and examines various types of FL. We focuses on the critical role of privacy preservation in FL in Sect. 4. In Sect. 5, we explore FL Applications, presenting a comprehensive overview of the potential uses of FL in different domains of smart healthcare. Finally, Sect. 6 concludes the paper.

2 Motivations of Using FL in Smart Healthcare

In this section, we explain the key motivations of the use of FL in smart healthcare [13].

2.1 Motivations

To emphasize the motivations, we will first explore the shortcomings of current healthcare systems, followed by a discussion of the advantages that FL can offer to the healthcare industry.

Challenges in Current Smart Healthcare Systems.

- **Privacy Issues:** Smart healthcare systems using AI often share data through the cloud, exposing information to privacy risks and unauthorized modifications.
- **Data Shortages at Medical Sites:** The lack of data limits the effectiveness of AI in healthcare, and sharing is hindered by privacy concerns and policies.
- **Limited performance of AI models:** Training on a single medical site often suffers from low accuracy due to imbalanced data, and GANs (Generative Adversarial Networks) do not always ensure the necessary diversity for effective application in healthcare.
- **High costs of health data processing:** Transferring data to the cloud results in significant latency and complicates the design of medical devices.

2.2 Benefits of FL in Smart Healthcare

FL offers several significant advantages for advancing smart healthcare due to its innovative operational approach:

- **Enhanced data privacy:** FL improves privacy by sending only local updates to the central server, reducing the risk of sensitive information leakage and promoting safer smart healthcare systems.

- **Balance between accuracy and utility:** FL enhances data privacy while maintaining strong generalization and improving the scalability of smart healthcare systems.
- **Reduction of health data training costs:** FL avoids large-scale data transfer to the server, thereby reducing latency, energy consumption, and the risk of congestion in healthcare networks.

3 Data Partitioning

FL provides the ability to train AI models without the need to share training data. Unlike traditional AI methods that rely on centralized data, current ML algorithms embrace a decentralized approach. This allows models to be collaboratively trained across a multitude of devices, such as smartphones, laptops, or private servers.

3.1 Key Principle

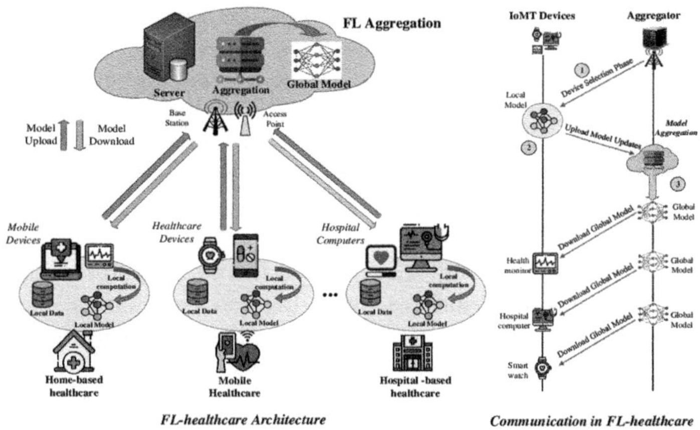

Fig. 1. The network architecture and communication process for FL-smart healthcare.

To understand how FL functions in smart healthcare, let's review the key steps of the process, as shown in Fig. 1 [13]. The process involves several key stages:

Client Selection and Initialization: The central server selects a specific medical task and the clients (hospitals or devices) that will participate in the FL process.

Local Training and Updates: Each client trains a local model on its data and then sends the updates back to the central server.

Aggregation and Distribution: The central server combines the received updates, updates the global model, and sends it back to the clients. This cycle repeats until the desired goal is achieved.

3.2 Types of FL for Smart Healthcare

FL is divided into three categories: horizontal, vertical, and hybrid, as described by Himeur et al. [14]. This paper explores these types, adapting them specifically to the field of smart healthcare.

Horizontal FL (HFL): as portrayed in Fig. 2 [14], in the context of HFL applied to smart healthcare, users collaborate to develop a shared global model by leveraging their respective datasets. Although these datasets share the same features, their samplings differ. To illustrate this process, let's imagine n users each recording their own voice saying the same phrase on their smartphones. Each user i, where $i \in \{1, 2, ..., n\}$, has their own local dataset, denoted as D_i, consisting of inputs X_i and their corresponding labels Y_i. The objective is to train a global model f capable of effectively generalizing across all users datasets. The training process in HFL typically involves the following steps: (1) The process starts with the initialization of local model parameters θ_i for each user. (2) Then, each user conducts local training on their vocal data to detect speech disorders, optimizing a local loss function $L_i(\theta_i)$ at each iteration. (3) After this training, users encrypt their gradients computed with respect to their local loss functions and transmit them to a central server for processing. (4) The trusted aggregator or central server performs secure aggregation on the encrypted gradients received from the users. Secure aggregation techniques, such as secure multiparty computation (SMPC) or homomorphic encryption..., are employed to aggregate the gradients while maintaining privacy. The aggregator can compute the average of the encrypted gradients or perform more sophisticated aggregation schemes, such as averaging the parameters, of θ_i to obtain the global model parameters θ. (5) Once the secure aggregation is completed, the aggregated gradients are decrypted by the trusted aggregator or central server. The global model parameters θ are then updated using the decrypted aggregated gradients. This update ensures that the global model benefits from the collective knowledge of all users while preserving the privacy of individual data.

Mathematically, the objective of HFL typically aims to minimize a global loss function $L(\theta)$ over all users' datasets, which can be formulated as:

$$L(\theta) = \sum_{i=1}^{N} L_i(\theta_i) + \lambda.R(\theta), \tag{1}$$

where $\sum_{i=1}^{N} L_i(\theta_i)$ represents the sum of local loss functions over all users' datasets, λ is a regularization parameter, and $R(\theta)$ represents a regularization term that helps prevent overfitting and encourages model simplicity.

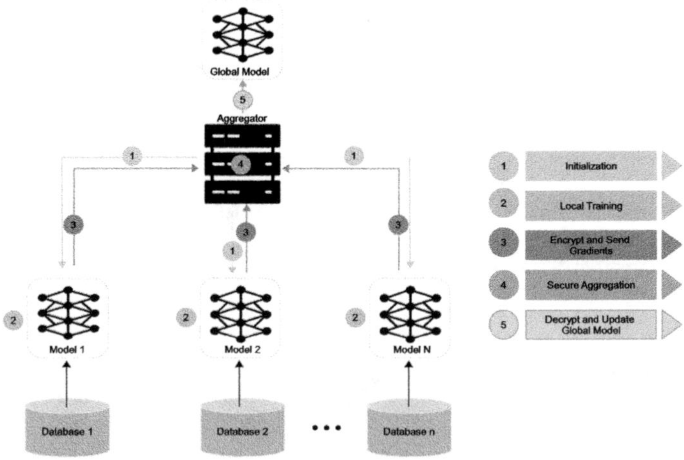

Fig. 2. Architecture of HFL.

Vertical FL (VFL): as illustrated in Fig. 3 [14], VFL trains ML models on datasets having the same sample space, but different feature spaces. In this context, techniques that rely on entity alignment can be used in combination with encryption to overcome the problem of data sample overlapping at distributed clients during the local training. With VFL different clients can cooperatively train their models without sharing their data (that differ in features), by only sharing their predictions for the samples they have in common.

Let's consider a scenario with two parties, party A and party B. Party A holds a dataset with features representing historical medical records in hospitals, while party B holds a dataset with features representing healthcare costs at the insurance company. The objective is to jointly train a model that leverages complementary information from both parties' datasets without sharing their raw data. The training process in VFL typically involves the following steps: (1) Preprocess the datasets of party A and party B to ensure compatibility and feature alignment, such as data transformation, feature mapping, etc. (2) Initialize the parameters of the model θ_A, θ_B, and θ for local and global models. (3) Each party independently conducts local training on its own dataset using its local model parameters θ_A and θ_B. Party A uses features from historical medical records in hospitals, while party B uses features from healthcare costs at the insurance company. Local loss functions $L_A(\theta)$ and $L_B(\theta)$ are optimized to update local model parameters and reduce divergences between the local model predictions and corresponding labels in each party's dataset. Steps (4), (5), and (6) are similar to steps (3), (4), and (5) in HFL. The mathematical formulation of $L(\theta)$ in VFL, for N parties, remains the same as in the case of HFL (Eq. 1).

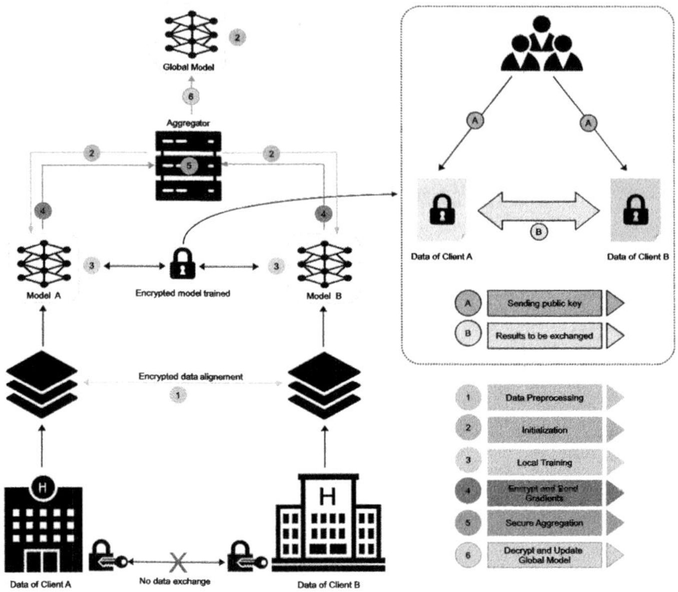

Fig. 3. Architecture of VFL.

Federated Transfer Learning (FTL): handles datasets with limited overlap both in the sample space and in the feature spaces, as shown in Fig. 4 [14]. Unlike VFL systems the number of common samples and thus labels that can be shared across the models is small thus limiting their reusability across the different models. A common transfer space is used for mapping the different feature spaces and exchanging knowledge. Specifically, adopting transfer learning methods helps calculate feature values from distinct feature spaces, which is utilized for training local datasets.

Let's consider there are N parties involved in FTL, where each party i (e.g., a hospital or an insurance company) has its own local dataset denoted as D_i. Each local dataset D_i represents patient medical records from a particular institution, including information such as age, gender, medical history, test results, etc., and each party i has a specific task denoted as T_i, for example, diagnosing cardiovascular diseases from patient medical records. The specific task T_i involves training a model capable of predicting the risks of cardiovascular diseases among patients in institution i.

By combining these examples, we can see how different parties may have specific datasets and tasks but can collaborate to train a global model through Federated Transfer Learning (FTL).

The training process in FTL typically involves the following steps: (1) Each party i performs local training on its own D_i and task-specific model parameters θ_i. During this step, the local model optimizes a task-specific loss function $L_i(\theta_i)$ to minimize the discrepancy between the local model's predictions and the corresponding labels in D_i. (2) The parties communicate and align their models (shared layers or parameters) to create a joint model architecture. (3) This transfer of knowledge, from party i to the global model noted as $Ki(\theta)$, can occur through parameter sharing or feature extraction. steps (4) and (5) remain the same as HFL. The global loss function $L(\theta)$ in FTL typically includes a combination of task-specific loss terms and transfer-related terms.

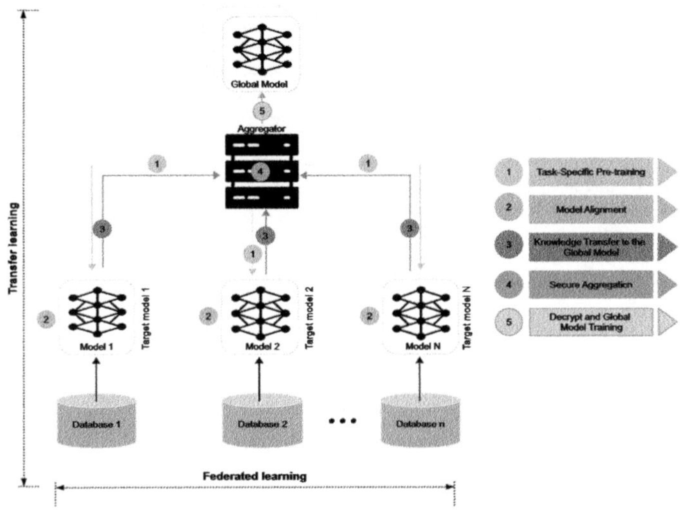

Fig. 4. Architecture of FTL.

4 Privacy Mechanisms

A key aspect of FL is that participants can keep their data locally without sharing it with others. Only model-related information is exchanged to train the global model. However, even this model information can potentially expose private data.

4.1 Aggregation Models

Various aggregation techniques in FL have been proposed in the literature. We will now explore some of the most relevant methods, specifically tailored for FL in smart healthcare applications. These methods include:

Averaging Aggregation: There are many federated averaging (FedAvg) aggregation strategies in the literature. FedAvg algorithm is thoroughly detailed in [15]. In Table 1, these aggregation algorithms are summarized, showing their main concept, along with an example in the field of smart healthcare, as well as their advantages and disadvantages.

Table 1. Comparative summary of the main FL aggregation techniques for Smart Healthcare.

Aggregation Technique	Main Concept	Potential Applications in Smart Healthcare	Advantages	Disadvantages
FedAvg	Clients update their models locally and send weights to a central server.	Patient monitoring models based on connected medical devices.	Simple and effective.	Sensitive to outliers and non-IID (Non-Independent and Identically Distributed) data.
FedProx	Adds a proximal term to local loss to reduce divergence.	Ensuring consistent updates in diagnostic models.	Improves convergence on heterogeneous data.	Device capability differences not addressed.
FedPAQ	Clients perform multiple local updates before sending to the server.	Personalized models for chronic disease management, e.g., dietary recommendations.	Tailors models to user needs.	Complexity in averaging models.
Turbo-Aggregate	Divides clients into groups with secure additive sharing.	Aggregating data from multiple clinical research sites for disease prediction.	Good for wireless networks and handles user dropouts.	Doesn't easily add new users; requires a self-configuring protocol.
FedMA	Adjusts global model size by considering neuron permutation.	Aggregating diagnostic models, e.g., cancer detection.	Adapts model size to data heterogeneity.	Vulnerable to model poisoning attacks.
HierFAVG	Uses hierarchical aggregation via edge and cloud servers.	Enhances medical data security with edge-cloud architecture.	Optimizes model exchange across edge-cloud systems.	Susceptible to stragglers and dropouts.

Progressive Fourier Aggregation (PFA): To address the retrogress problem in real-world medical FL, PFA [16] is employed at the server level, gradually integrating low-frequency client models with high-frequency ones. A Deputy-Enhanced Transfer (DET) strategy is implemented on the client side to effectively transfer global knowledge to the personalized local model. The Fast Fourier Transform (FFT) is used to obtain amplitude and phase maps of the parameters, while the Inverse Fourier Transform (IFFT) allows for the reverse conversion. In summary, FFT enhances the understanding of data structures, promoting personalized care.

FedGKT Aggregation: FL stands out for its ability to protect the confidentiality of users' data. However, edge nodes are often limited in terms of computing power. While the use of convolutional neural networks (CNNs) can optimize model performance, large models can overload these nodes and lead to high communication costs. To address these issues, the FedGKT algorithm [17] adopts a knowledge transfer approach that reduces the amount of data to be communicated. This mechanism requires 9 to 17 times fewer computing resources compared to FedAvg while maintaining comparable accuracy. Thus, FedGKT is particularly suited for smart healthcare systems, such as the development of mammography image analysis systems, where multiple hospitals can collaborate while preserving the confidentiality of patient information.

4.2 Secure Aggregation

The aggregator alone does not guarantee the security of the exchanged data, as it may be vulnerable to attacks from malicious aggregators or users. To mitigate these risks, it is necessary to employ SA techniques, such as differential privacy, trusted execution environment (TEE), anonymity, or cryptography. SA typically involves two main actors: Users and Aggregators. Users are the entities that provide the input data, while aggregators are responsible for combining this data to obtain an aggregated result.

SA Based on Differential Privacy: Differential privacy is a concept that theoretically explores whether a modification of a single data point in the training dataset of an algorithm leads to statistically insignificant changes in the algorithm's output. To achieve secure aggregation using differential privacy mechanisms, users introduce statistical noise into their trained models [18]. By integrating differential privacy into smart healthcare, we can ensure the security of patient data while allowing for accurate and relevant analyses.

SA Based on TEE: Another method for ensuring secure data aggregation is the use of a TEE on the server [19]. A TEE is a protected area within the main processor designed to maintain the confidentiality and integrity of the data being processed. By performing data aggregation within the TEE, the security of the data is ensured, preventing unauthorized access or modification during the processing. The TEE enables secure computations on sensitive medical data during training, thereby preserving both their integrity and confidentiality. Furthermore, it facilitates real-time analysis through connected medical devices, which is particularly relevant in the field of smart healthcare, while securing communications with servers.

SA Based on Anonymity: This approach employs principles of anonymous communication [20]. Also known as "secure shuffling," it involves dividing client data into multiple shares, which are then sent anonymously to the aggregator [19]. This process ensures that the server cannot identify the origin of the received shares and thus cannot reconstruct the users' data. In healthcare, anonymization allows for the reporting of adverse treatment effects without compromising confidentiality. Medical data is fragmented and anonymized before being sent for analysis, which facilitates trend detection while safeguarding personal information. This method promotes participation, improves care monitoring, and maintains user trust.

SA Based on Cryptography: This method uses cryptographic techniques to ensure that sensitive data remains confidential throughout the aggregation process. Cryptography encompasses various methods designed to secure data, including homomorphic encryption and secure multiparty computation (SMPC).

Homomorphic encryption is a cryptographic method that allows operations to be performed on encrypted data without needing to decrypt it. This technique ensures that sensitive information remains protected throughout processing and analysis. In the context of healthcare, homomorphic encryption is particularly useful for maintaining the confidentiality of medical data during collaborations between institutions. For example, recent work shows how homomorphic encryption is used to analyze neuroimaging data while preserving its confidentiality [21], or to combine this technique with blockchain to secure federated model aggregation for medical images [22].

SMPC is another cryptographic technique that allows multiple entities to perform calculations on distributed data while maintaining its confidentiality [23,24]. In the context of FL in healthcare, SMPC enables multiple medical institutions to collaborate on training predictive models without directly sharing their sensitive data, such as patient medical records. Each institution locally encrypts its data and then participates in a secure computation, allowing for the creation of a global model without ever revealing personal information.

5 FL Applications in Healthcare

In this section, we explore the emerging applications of FL in smart healthcare.

5.1 FL Applications for Medical Image Processing

Medical imaging revolutionizes clinical practice by enabling detailed visualization of the human body, but secure data sharing remains a significant challenge. Due to growing privacy concerns, it is difficult to implement artificial intelligent based medical data imaging by consolidating data from different medical institutions into a centralized entity. FL emerges as a promising solution by allowing model training on data from multiple sources without disclosing patients' private information. This approach is crucial for the rapid and accurate reconstruction of magnetic resonance imaging in critical clinical applications. FL algorithms offer a viable path by preserving privacy while improving the accuracy of brain tumor segmentation and MRI classification, thereby facilitating early and precise disease diagnosis. In summary, FL represents a significant advancement in medical image analysis by ensuring the protection of patients' personal data.

The application of FL for brain imaging is described in [25]. Each client locally trains a deep neural network (DNN) model on brain tumor data and then shares the updated parameters for central aggregation. However, to reduce the risk of data leakage during communication, differential privacy is employed, adding noise to the updates of the local models. This method, validated on the BraTS 2018 dataset, ensures effective data protection while maintaining model performance within a secure collaborative framework.

Yan et al.[26] present an innovative FL approach to manage variations in medical image data from multiple decentralized sources. Their method reduces inter-client differences by transforming images into a common space using a privacy-preserving generative adversarial network, named PPWGAN-GP. Subsequently, a modified CycleGAN is deployed for each client, allowing the transfer of raw images to the target image space defined by shared synthesized images. This technique effectively addresses the problem of inter-client variations while ensuring data privacy. The results show a significant improvement in the accuracy of classification models, demonstrating the effectiveness of this approach for medical image processing in a decentralized environment.

Qayyum et al. [27] present a collaborative FL framework for COVID-19 diagnosis using multimodal data such as X-rays and ultrasounds, processed at the edge. The CFL approach allows remote healthcare centers to collaborate effectively without sharing sensitive patient data, thus preserving confidentiality while improving diagnostic accuracy. The authors also demonstrate how this method mitigates data distribution divergence issues observed in conventional FL, highlighting both the advantages and challenges of deploying machine learning models at the edge in medical applications.

Hashmani et al. [28] propose an intelligent system based on adaptive FL for skin disease detection. This system aims to enhance dermatologists' diagnostic capabilities using an intelligent dermoscopy device. By integrating deep learning techniques and a continuous adaptation mechanism, the model can accurately classify dermoscopy images and learn new features from new samples. The results demonstrate satisfactory classification performance, which is crucial for clinical applications.

Lee et al. [29] used ultrasound image analysis in conjunction with FL to assess whether thyroid nodules were benign or malignant. This approach allowed them to classify the nodules using data from multiple sources without directly sharing sensitive medical data, thereby enhancing patient privacy protection while optimizing the performance of the learning algorithm.

Linardos et al. [30] developed a method for cardiovascular magnetic resonance imaging using a FL framework specifically designed for diagnosing hypertrophic cardiomyopathy. They employed a 3D convolutional neural network (3D-CNN) model, which was pre-trained for action recognition. Additionally, prior shape information was incorporated into the 3D-CNN using two distinct methods and four data augmentation strategies. This approach was tested and validated on the Automatic Cardiac Diagnosis Challenge (ACDC) dataset.

Dayan et al. [31] developed a FL model to predict future oxygen needs in patients with COVID-19. This model, named EXAM, was trained using data from 20 institutes worldwide. Based on the COV-2 clinical decision support model, EXAM utilizes chest X-rays, laboratory data, and vital signs to make its predictions. This approach allows for forecasting oxygen needs without requiring the direct sharing of sensitive medical data.

5.2 Managing EHRs Using FL

Traditional artificial intelligence techniques pose privacy issues, potentially leading to the leakage of sensitive data during analysis. Although the patient's name is removed from health records for security reasons, this measure remains insufficient. Indeed, hospitals and agencies often collaborate to find solutions to new diseases and commonly share a single central server where patient data is stored, which amplifies privacy risks. In this context, techniques based on FL offer a more reliable and intelligent solution for managing EHR. FL helps preserve patient data privacy, even when multiple entities cooperate, by training AI models directly on local devices without centralizing sensitive data.

An approach based on FL is proposed in [32] by Brisimi et al. to predict hospitalizations of patients diagnosed with heart diseases using their historical EHRs. Health data from patients' smartphones and distributed hospitals are locally trained, considering demographic information such as age, gender, and physical characteristics. Subsequently, the trained model parameters are aggregated by a cloud server to build a unified prediction model based on a global support vector machine (SVM) classifier. The goal is to predict future hospitalizations due to heart diseases for hospital resource management, while preserving data privacy.

One such example was presented in [33] by Hao et al.in which privacy and optimal resource usage-based protocol schemes were presented for FL to analyze EHR. To further enhance the privacy in local model parameters, perturbation of training data is performed to prevent memorization attacks. The main advantage of this technique is that even if the adversary can get information about perturbations in EHR, the original information will remain safe and secured.

Choudhury et al. [34] propose a method for managing EHRs using FL combined with differential privacy. Their approach allows for training a global model while keeping data localized at each site, adding artificial noise to model updates to protect sensitive information. This solution ensures high data protection while facilitating secure collaboration between health institutions.

5.3 Drug Discovery and Development via FL

Drug development relies on analyzing large sets of biomedical data. Centralizing this data raises confidentiality and security issues, which can limit collaboration between laboratories and institutions. FL offers an innovative solution by enabling the training of artificial intelligence models on locally distributed data without the need for centralization. This approach promotes inter-institutional collaboration while ensuring the protection of sensitive data. Additionally, it accelerates the discovery of new molecules and enhances drug design through more efficient and precise analysis of pharmacological interactions and patient characteristics, all while adhering to current confidentiality standards.

The QSAR (Quantitative Structure-Activity Relationship) analysis is an essential tool in drug discovery, allowing for the prediction of the activity of compounds. Chen et al. [35] present the application of HFL as an innovative and privacy-preserving solution for QSAR analysis.

The MELLODDY project, detailed by Heyndrickx et al. [36], demonstrates the use of FL to improve QSAR models while preserving data confidentiality, in collaboration with ten pharmaceutical companies. This approach illustrates how FL can transform drug discovery and development by enabling large-scale collaborative analysis without compromising sensitive information. The practical application of MELLODDY has been demonstrated [37].

Hanser [38] explores the use of FL in molecular discovery, emphasizing its potential to improve collaboration between institutions while protecting data privacy. A key challenge is the non-identical (non-IID) distribution of QSAR data, caused by differences in chemical properties and activities across organizations. This makes it harder for models to generalize across institutions. Hanser highlights the need for methods that can address this non-IID data to enhance model performance in collaborative drug discovery.

5.4 Patient Monitoring and Remote Healthcare

The increasing demand for smart solutions in remote healthcare is driving a shift from hospital-based care to home-based care. Home patient monitoring can be improved with technologies like FL, which allows for building a global model while keeping user data stored locally, reducing the risk of data leaks. Using connected devices IoMT (Internet of Medical Things) to tailor prediction models helps enhance accuracy and effectively manages issues with imbalanced and non-IID data.

Chen et al. [39] have developed FedHealth, a transfer learning federated framework for portable healthcare. This system aggregates data through FL and develops personalized models using transfer learning. Tests conducted on wearable devices for activity recognition and a diagnostic application for Parkinson's disease have demonstrated that FedHealth can provide accurate and personalized healthcare while ensuring data privacy and security. It is a general framework that is adaptable to numerous healthcare applications.

Xu et al. [40] propose a method for detecting depression using health data collected from mobile phones. The authors utilize keystroke recordings and other data from various devices to predict depression. The approach achieves high accuracy even in non-IID data environments while ensuring the security and privacy of personal information.

Liu et al. [41] present an FL model for smart healthcare systems focused on addressing confidentiality and efficiency in medical data collected through IoT devices. Their solution employs a dynamic secret sharing method with advanced cryptographic techniques and a user authentication system to safeguard local data. Furthermore, this model supports user participation flexibility, allowing users to join or leave the learning process without compromising model integrity.

The Table 2 summarizes the different FL approaches applied to smart healthcare presented in this paper. It provides a description of each approach, along with their respective advantages and limitations.

Table 2. Summary of FL Approaches in Smart Healthcare.

Category	Examples	Description	References	Advantages	Disadvantages
Medical Image Analysis	Brain tumor segmentation	FL with differential privacy for segmenting tumors, enhancing model accuracy while safeguarding data.	Li et al. [25]	Enhances privacy, improves segmentation accuracy.	Compatibility issues between data from sites; risk of data leakage.
	Cardiac pathologies detection	FL for diagnosing cardiac pathologies from MRIs or CXR images.	Linardos et al. [30]	Preserves confidentiality; diverse data access.	Increased complexity in managing models; difficulty aggregating data.
	Skin disease detection	Adaptive FL system for skin disease detection using dermoscopy images.	Hashmani et al. [28]	Enhances diagnostics; continuous model adaptation.	Misinterpretation risks; challenges in cloud deployment.
	Thyroid diagnosis	Ultrasound analysis via FL to determine nodule benignity, enhancing privacy.	Lee et al. [29]	Improves patient privacy; optimizes performance.	More data needed for better performance; imbalance issues.
	Cancer detection	FL using PPWGAN-GP and CycleGAN to align data from institutions.	Yan et al. [26]	Better diagnostic accuracy; preserves data privacy.	High computational complexity; heterogeneous data management challenges.
	COVID diagnosis	FL to analyze ultrasound images for COVID-19 detection.	Dayan et al. [31]	Enhanced detection capabilities; preserves privacy.	Requires large, diverse datasets; handling data variations is complex.
	COVID-19 framework	CFL-based diagnosing framework using X-rays and ultrasounds at the edge.	Qayyum et al. [27]	Maintains confidentiality; enhances accuracy via data integration.	Edge deployment complexity; quality variability; data heterogeneity issues.
Managing EHRs	Heart disease hospitalization prediction	FL for predicting hospitalizations from EHRs, considering demographics.	Brisimi et al. [32]	Predicts hospitalizations while preserving privacy; diverse data usage.	Dependence on local data accuracy; aggregation challenges.
	Privacy-enhanced EHR analysis	FL with perturbation techniques to analyze EHRs, preventing memorization attacks.	Hao et al. [33]	Enhances privacy; protects against attacks.	Implementation complexity; performance trade-offs.
	FL for sensitive health data	FL framework for training on distributed EHRs, ensuring privacy.	Choudhury et al. [34]	High data privacy; collaborative training without centralization.	Differential privacy implementation complexity; local data quality dependency.
Drug Discovery and Devel-opment	FL-QSAR	Privacy-preserving FL for QSAR analysis, improving drug discovery performance.	Chen et al. [35]	Enhances collaboration; achieves results similar to traditional methods.	Complexity in FL implementation.
	QSAR improvement with MELLODDY	FL to enhance QSAR models in drug discovery with major pharma collaboration.	Heyndrickx et al. [36]	Improves prediction; preserves confidentiality; large-scale analysis.	Coordination complexity among entities; potential FL setup issues.
	FL for QSAR Data Analysis	Integrates FL to manage non-IID distribution of QSAR data, improving performance.	Hanser [38]	Enhances model generalization; facilitates collaboration.	Challenges in handling non-IID data; ensuring accuracy across datasets.
Patient Monitoring and Remote Healthcare	FedHealth	Federated transfer learning framework for personalized healthcare using FL.	Chen et al. [39]	Accurate, personalized care; maintains data privacy.	Needs evaluation of communication costs and training complexity.
	Depression detection	FL using mobile health data, including keystroke recordings, to predict depression.	Xu et al. [40]	High accuracy in non-IID data; ensures information confidentiality.	Relies on mobile health data quality; diverse sources needed for optimality.
	FL for IoT-based healthcare systems	FL scheme based on dynamic secret sharing for improving data privacy and system efficiency.	Liu et al. [41]	Improves data privacy; reduces computational overhead; supports dynamic user participation.	Requires robust user authentication; potential integration challenges when users join or leave the learning process.

6 Conclusion

In this paper, we have presented an in-depth study of the use of FL in the context of smart healthcare systems. In particular, our study has focused on three key aspects of FL in healthcare, notably data partitioning strategies, privacy preservation mechanisms, and applications in the healthcare context. We have discussed different approaches to data partitioning, highlighting their importance for the effectiveness of FL models in decentralized healthcare systems. Privacy preservation mechanisms have been analyzed, underlining their crucial role in protecting patient's confidential data. We have proposed a taxonomy to classify existing FL-based approaches in the healthcare field. This taxonomy is structured around four main categories, namely FL applications for medical image processing, FL-enabled management of EHRs, FL-enabled drug discovery and development, and patient monitoring and remote Healthcare. We've systematically analyzed the different approaches proposed in the literature, highlighting their strengths and limitations.

An interesting perspective would be to conduct thorough experimental studies to quantitatively compare the different aggregation methods and assess their performance in real-world scenarios. Additionally, it would be valuable to explore the integration of FL with distributed computing architectures such as Edge-Fog-Cloud Computing. This integration could significantly improve the efficiency, latency, and scalability of learning processes in decentralized healthcare systems, enabling more responsive and adaptive healthcare solutions.

Acknowledgments. This work is supported by GREYC laboratory, Normandy University, France. It was performed in collaboration with LIMED Laboratory, Bejaia University, Algeria, under their tutelage DGRSDT.

References

1. Aceto, G., Persico, V., Pescape, A.: Industry and health: internet of things, big data, and cloud computing for healthcare. J. Ind. Inf. Integr. **18**, 100129 (2020)
2. Said, D., Elloumi, M., Khoukhi, L.: Cyber-attack on p2p energy transaction between connected electric vehicles: a false data injection detection based machine learning model. IEEE Access **10**, 640–647 (2022)
3. Rieke, N., Hancox, J., Li, W., Milletarì, F., Roth, H.R., Albarqouni, S., Cardoso, M.J.: The future of digital health with federated learning. NPJ Digital Med. **3**(1), 1–7 (2020)
4. Moudoud, H., Cherkaoui, S., Khoukhi, L.: Towards a secure and reliable federated learning using blockchain, pp. 1–6 (2021)
5. Kaissis, G., Makowski, M.R., Rucert, D., Braren, R.F.: Secure, privacy-preserving and federated machine learning in medical imaging. Nat. Mach. Intell. **2**(6), 305–311 (2020)
6. Xu, J., Glicksberg, B.S., Su, C., Walker, P., Bian, J., Wang, F.: Federated learning for healthcare informatics. J. Healthc. Inf. Res. **5**, 1–19 (2021)
7. Kaissis, G., et al.: End-to-end privacy preserving deep learning on multi-institutional medical imaging. Nat. Mach. Intell. **3**(6), 473–484 (2021)
8. Roth, H.R., et al.: Federated learning for breast density classification: a real-world implementation. In: Domain Adaptation and Representation Transfer, and Distributed and Collaborative Learning, pp. 3181–191. Springer (2020)
9. Abou El Houda, Z., Hafid, A.S., Khoukhi, L.: A novel machine learning framework for advanced attack detection using SDN, pp. 1–6 (2021)
10. Liu, D., Khoukhi, L., Hafid, A.: Data offloading in mobile cloud computing: a Markov decision process approach, pp. 1–6 (2017)
11. Ali, M., Naeem, F., Tariq, M., Kaddoum, G.: Federated learning for privacy preservation in smart healthcare systems: a comprehensive survey. IEEE J. Biomed. Health Inform. **27**(2), 778–789 (2022)

12. Dasaradharami Reddy, K., Gadekallu, T.R.: A comprehensive survey on federated learning techniques for healthcare informatics. Comput. Intell. Neurosci. **2023**(1), 8393990 (2023). https://onlinelibrary.wiley.com/doi/abs/10.1155/2023/8393990
13. Nguyen, D.C. et al.: Federated learning for smart healthcare: a survey (2021). https://arxiv.org/abs/2111.08834
14. Himeur, Y., et al.: Federated learning for computer vision. arXiv preprint: arXiv:2308.13558 (2023)
15. Wahab, O.A., Mourad, A., Otrok, H., Taleb, T.: Federated machine learning: survey, multi-level classification, desirable criteria and future directions in communication and networking systems. IEEE Commun. Surv. Tutorials **23**(2), 1342–1397 (2021)
16. Chen, Z., Zhu, M., Yang, C., Yuan, Y.: Personalized retrogress-resilient framework for real-world medical federated learning. In: International Conference on Medical Image Computing and Computer-Assisted Intervention, pp. 347–356. Springer (2021)
17. Chaoyang, H. E., Annavaram, M., Avestimehr, S.: Group knowledge transfer: collaborative training of large CNNs on the edge (2020)
18. Moustafa, A., Asad, M., Shaukat, S., Norta, A.: PPCSA: partial participation-based compressed and secure aggregation in federated learning, LNNS 226, pp. 345–357 (2021)
19. Zhang, Y., Wang, Z., Cao, J., Hou, R., Meng, D.: ShuffleFL: gradient-preserving federated learning using trusted execution environment, ser. CF '21, pp. 161–168. Association for Computing Machinery, New York (2021). https://doi.org/10.1145/3457388.3458665
20. Ishai, Y., Kushilevitz, E., Ostrovsky, R., Sahai, A.: Cryptography from anonymity. In: Proceedings of the 47th Annual IEEE Symposium on Foundations of Computer Science, ser. FOCS '06, pp. 239–248. IEEE Computer Society, USA (2006). https://doi.org/10.1109/FOCS.2006.25
21. Stripelis, D., et al.: Secure neuroimaging analysis using federated learning with homomorphic encryption (2021). https://arxiv.org/abs/2108.03437
22. Kumar, R., et al.: Blockchain and homomorphic encryption based privacy-preserving model aggregation for medical images. Comput. Med. Imag. Graph. **102**, 102139 (2022)
23. Zhu, H.: On the relationship between (secure) multi-party computation and (secure) federated learning. arXiv preprint: arXiv:2008.02609 (2020)
24. Byrd, D., Polychroniadou, A.: Differentially private secure multiparty computation for federated learning in financial applications. In: Proceedings of the First ACM International Conference on AI in Finance, pp. 1–9 (2020)
25. Li, W., et al.: Privacy-preserving federated brain Tumour segmentation (2019)
26. Yan, Z., Wicaksana, J., Wang, Z., Yang, X., Cheng, K.T.: Variation-aware federated learning with multi-source decentralized medical image data. IEEE J. Biomed. Health Inf., 1–1 (2020)
27. Qayyum, A., Ahmad, K., Ahsan, M.A., Al-Fuqaha, A., Qadir, J.: Collaborative federated learning for healthcare: multi-modal COVID-19 diagnosis at the edge (2021)
28. Hashmani, M.A., Jameel, S.M., Rizvi, S.S.H., Shukla, S.: An adaptive federated machine learning-based intelligent system for skin disease detection: a step toward an intelligent Dermoscopy device. Appl. Sci. (2021). https://api.semanticscholar.org/CorpusID:233796931

29. Lee, H., et al.: Federated learning for thyroid ultrasound image analysis to protect personal information: validation study in a real health care environment. JMIR Med. Inform. **9**(5), e25869 (2021)
30. Linardos, A., Kushibar, K., Walsh, S., Gkontra, P., Lekadir, K.: Federated learning for multi-center imaging diagnostics: a study in cardiovascular disease (2021)
31. Dayan, I., et al.: Federated learning for predicting clinical outcomes in patients with COVID-19. Nat. Med. **27**(10), 1735–1743 (2021)
32. Brisimi, T.S., Chen, R., Mela, T., Olshevsky, A., Paschalidis, I.C., Shi, W.: Federated learning of predictive models from federated electronic health records. Int. J. Med. Inf. **112**, 59–67 (2018). https://api.semanticscholar.org/CorpusID:3679574
33. Hao, M., Li, H., Xu, G., Liu, Z., Chen, Z.: Privacy-aware and resource-saving collaborative learning for healthcare in cloud computing, pp. 1–6 (2020)
34. Choudhury, O., et al.: Differential privacy-enabled federated learning for sensitive health data (2020). https://arxiv.org/abs/1910.02578
35. Chen, S., Xue, D., Chuai, G., Yang, Q., Liu, Q.: FL-QSAR: a federated learning-based QSAR prototype for collaborative drug discovery. Bioinformatics **36**, 5492–5498 (2021)
36. Heyndrickx, W., et al.: MELLODDY: cross pharma federated learning at unprecedented scale unlocks benefits in QSAR without compromising proprietary information (2022)
37. Oldenhof, M., et al.: Industry-scale orchestrated federated learning for drug discovery. In: Proceedings of the AAAI Conference on Artificial Intelligence, vol. 37, pp. 576–584 (2023)
38. Hanser, T.: Federated learning for molecular discovery. Curr. Opin. Struct. Biol. **79**, 102545 (2023)
39. Chen, Y., Qin, X., Wang, J., Yu, C., Gao, W.: FedHealth: a federated transfer learning framework for wearable healthcare (2019)
40. Xu, X., et al.: Federated depression detection from multi-sourcemobile health data (2021)
41. Liu, W., Zhang, Y., Han, G., Cao, J., Cui, H., Zheng, D.: Secure and efficient smart healthcare system based on federated learning. Wirel. Commun. Mob. Comput. **2023**, 8017489 (2023)

Complex Communication Networks Management with Distributed AI: Challenges and Open Issues

Christina Alhachem[1(✉)], Mounir Kellil[1], and Abdelmadjid Bouabdallah[2]

[1] Université Paris-Saclay, CEA, List, 91120 Palaiseau, France
christina.alhachem@cea.fr

[2] Université de Technologie de CompièGne, HEUDIASYC UMR 7253, CompièGne, France

Abstract. The rapid evolution of communication networks, particularly with the introduction of 5G and the anticipated arrival of 6G, has introduced new complexities in managing traffic, routing, and resource allocation. Distributed Artificial Intelligence (DAI) is emerging as a promising solution to address challenges related to scalability, adaptability, and performance in these dynamic environments. With the unpredictable nature of network traffic patterns and the dynamic infrastructure of modern networks, effective network management is crucial for ensuring optimal resource utilization and preventing congestion. This is essential to maintain high performance, reliability, and scalability in today's communication systems.

This paper explores the application of AI techniques in network management, with a focus on key areas such as congestion control, routing management, and traffic prediction. By examining both centralized and distributed AI approaches—such as Multi-Agent Reinforcement Learning (MARL) —it highlights their potential to enhance network efficiency, improve latency, increase throughput, and reduce packet loss. The paper also addresses the limitations of current methods, while discussing potential future directions for AI- driven solutions in large-scale, real-time network operations.

Keywords: Network management · complex communication networks · congestion control · traffic prediction · routing management · machine learning · distributed artificial intelligence

1 Introduction

The rapid growth of IoT devices and applications, combined with the evolving needs of network operators and end-users, presents significant challenges in managing modern communication networks. These networks must now accommodate increasing demands, including handling substantial data loads, providing greater bandwidth, achieving ultra-low latency, and maintaining energy efficiency.

In complex communication environments, high processing demands, unpredictable traffic patterns, and evolving network topologies strain resources, making efficient network management critical for preventing congestion and maintaining performance.

F. Hacène et al. (Eds.): MLN 2024, LNCS 15540, pp. 32–54, 2026.
https://doi.org/10.1007/978-3-032-00552-6_3

Dynamic adaptation is essential to ensure optimal Quality of Service (QoS) metrics—such as latency, bandwidth, and energy efficiency—while handling massive traffic loads and ensuring real- time responsiveness. These factors are particularly crucial for delivering seamless, time- sensitive services in emerging technologies like 5G, 6G, IoT, and mobile edge computing. Traditional network management solutions often rely on static rules, which are inadequate for the dynamic and evolving nature of modern networks. This has created a need for more adaptive, intelligent approaches capable of managing the increasing complexity of communication networks.

The demand for ultra-low latency, high bandwidth, and consistent QoS requires innovative, flexible strategies that can dynamically adapt to changing conditions in real time.

Current techniques also struggle with scalability and responsiveness, underscoring the need for decentralized, real-time solutions to meet the challenges of today's communication networks. While there has been significant research on AI-driven network management, much of the focus has been on centralized approaches. In contrast, less attention has been paid to recent advancements in Distributed AI for network management. This paper seeks to address this gap by exploring the latest developments in Distributed AI techniques, such as Multi-Agent Reinforcement Learning (MARL), specifically in the context of managing complex communication networks.

This study examines various approaches to network management in complex environments, categorizing different strategies from the literature and evaluating their potential. We provide an in-depth analysis of current and emerging ML applications in communication networks, focusing on key areas such as congestion control, traffic prediction, and routing management. Section 2 introduces key concepts, including complex communication networks, network management, and the role of machine learning (ML) and distributed AI (DAI) in these contexts. Section 3 identifies the core challenges faced in network management, particularly as networks grow in scale and complexity. In Sect. 5, we analyze centralized ML solutions for network management, assessing their strengths and limitations. In Sect. 6, we transition to distributed AI approaches, exploring how these models address the limitations of centralized systems by enhancing scalability, adaptability, and real-time decision-making in dynamic networks. Throughout the paper, we highlight both the potential and the challenges of implementing these solutions, providing an evaluation of current trends and open issues in AI-driven network management.

2 Concepts

2.1 Complex Communication Networks

Complex communication networks refer to modern and advanced networks, such as 5G, 6G, and large-scale IoT systems, characterized by their dynamic topologies, diverse applications, and varying traffic patterns. These networks must accommodate a vast number of connected devices, provide high data throughput, and meet stringent Quality of Service (QoS) requirements, such as ultra-low latency and high reliability. The complexity arises from factors like the integration of heterogeneous technologies, the need for real-time data processing, and the management of fluctuating traffic loads caused by event-driven applications and services like VR/AR. These characteristics make managing and optimizing such networks a challenging task, requiring advanced and adaptive solutions.

These networks have been defined in various ways in the literature, emphasizing their irregular structure, dynamic behavior, and representation of real-world systems. Table 1 summarizes several definitions, highlighting the diversity in how these networks are understood.

2.2 Network Management

Building on the concept of complex communication networks, network management aims to ensure efficient, secure, and reliable operation within these dynamic environments. Network management entails a range of different activities and concepts, including congestion control, routing management, traffic prediction, and performance optimization.

In these complex networks, nodes can become overloaded due to intensive processing tasks such as traffic routing, data analysis, video encoding/decoding, and data encryption/decryption.

Moreover, network bandwidth often exhibits unpredictable or chaotic patterns driven by high- bandwidth-consuming applications like virtual reality (VR), augmented reality (AR), and mixed reality (XR). Consequently, the application of automated reasoning algorithms is increasingly necessary to allow dynamic adaptation to changes in user experience and Quality of Service (QoS).

Congestion control (CC) is vital for preventing network overload. It manages data flow to minimize packet loss and latency, ensuring high throughput and maintaining network stability even under heavy traffic conditions.

Routing management also plays a significant role in network management by determining optimal paths for data transmission. Traditionally, this involves predefined algorithms and protocols that can be proactive, reactive, or hybrid, each offering different strategies for effectively directing traffic through the network.

Traffic prediction complements these efforts by forecasting future network traffic patterns based on historical data and current conditions. By anticipating changes in traffic behavior, the network can adapt proactively, optimizing its performance and resource allocation. Together, these aspects form the foundation of managing complex communication networks.

Table 1. Definitions of Complex Networks in Literature.

Reference	Definition
[1]	Networks whose structure is irregular, complex and dynamically evolving in time
[2]	By definition, complex networks are networks with more complex architectures than classical random graphs with their "simple" Poissonian distributions of connections
[3]	Complex weblike structures describe a wide variety of systems of high technological and intellectual importance

(continued)

Table 1. (*continued*)

Reference	Definition
[4]	For instance, there may be more than one different type of vertex in a network, or more than one different type of edge. And vertices or edges may have a variety of properties, numerical or otherwise, associated with them

2.3 Machine Learning and Distributed AI

Conventional automated reasoning solutions fall short of meeting these rigorous demands because they rely heavily on predefined rules and static knowledge bases, which are inadequate for capturing the dynamic and evolving nature of modern complex networks. This necessitates the development of new, more adaptable approaches.

To address this, the automated network management paradigm is shifting towards the integration of Artificial Intelligence (AI) techniques, including Machine Learning (ML). This shift aims to optimize network management operations while dynamically adapting to changing network conditions and user requirements. Particularly in the realm of future mobile communication networks like 6G, the stringent QoS demands necessitate a comprehensive approach to an increasingly intricate communication environment. This environment is marked by exponential growth in network traffic due to applications, video streaming, VR/AR/XR; the extensive deployment of IoT devices; the proliferation of mobile edge computing; and the emergence of highly dynamic network topologies.

While numerous ML techniques have been proposed for network management, most rely on a centralized controller for network-wide monitoring and decision-making. Despite simplifying network operations, these centralized ML methods face significant limitations, such as lack of scalability, a single point of failure, limited fault tolerance, high latency, and restricted service coverage. Distributed ML approaches present a promising alternative to overcome these limitations.

A promising approach in Distributed AI is MARL, where multiple AI agents can either collaborate or operate independently to optimize network management. In MARL, agents gather data from their environment, learn over time through feedback (rewards) for actions that enhance network performance, and can adjust parameters such as bandwidth allocation or traffic routing in response to local conditions. Whether collaborating or acting independently, these agents collectively improve network efficiency without depending on a centralized controller, which enhances scalability and resilience to failures.

Figure 1 illustrates the structure of MARL within a 5G core network. Each agent, such as Agent 2, processes local state information, observes rewards, and takes action based on these inputs. These agents can communicate with each other and share learned knowledge to enhance overall network performance. By distributing decision-making across the network, MARL allows for dynamic adaptation to traffic conditions in large-scale, complex networks like 5G and 6G.

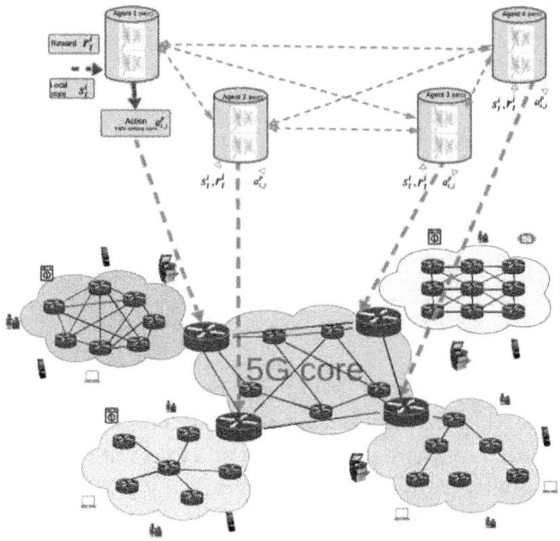

Fig. 1. Multi-Agent Reinforcement Learning (MARL) in a 5G Core Network.

3 Network Management Challenges

Managing complex communication networks for optimal performance and security presents numerous challenges that require strategic planning and meticulous oversight. These challenges are outlined below:

– Multi-layer interactions among interconnected nodes: The interactions between multiple layers in a network complicate management by making it harder to track and optimize performance across different layers. These interactions introduce dependencies and potential vulnerabilities that can affect the network's reliability and efficiency.
– Diversity of communication protocols: The wide range of communication protocols in use creates challenges for ensuring interoperability and optimal performance across the network.

Managing this diversity requires sophisticated configuration strategies to enable seamless communication between devices with varying capabilities and protocol requirements.

– Unpredictable traffic patterns: Sudden fluctuations in traffic, caused by high-demand applications or event-driven services, make it challenging to predict and manage network resources efficiently. The ability to anticipate and react to these unpredictable patterns is critical for maintaining optimal performance and avoiding congestion.
– Accommodating technological advancements: Integrating new technologies with legacy systems poses challenges in maintaining network functionality and performance. Balancing innovation with backward compatibility demands careful planning to ensure that new advancements enhance the network without disrupting existing services.
– Network congestion and control strategies: Network administrators must manage congestion and latency while balancing centralized and decentralized control. The

challenge is to develop strategies that mitigate congestion and reduce latency without compromising the benefits of either management approach, ensuring efficient and responsive network operations.

These challenges drive the need for new approaches, particularly those leveraging AI and machine learning.

4 Traditional Solutions

Traditional network management solutions have long relied on established techniques. Conventional adaptive congestion control (CC) protocols for Transmission Control Protocol (TCP) [5], ECN [6], and AQM [7] and its improvements such as BIC and CUBIC [8], are not well adapted to accommodate ever-evolving characteristics and requirements of communication networks, especially when it comes to considering unpredictable traffic patterns. This is particularly the case with event-driven applications (such as IoT notifications and alarms) and immersive scenarios like VR/AR/XR-based applications.

In routing management, traditional techniques have predominantly been rule-based, following predefined algorithms. Proactive protocols like Distance Vector Routing Protocol (DVRP) [10] and Open Shortest Path First (OSPF) [11] maintain constantly updated routing tables for immediate route availability but are less scalable due to the high overhead involved. Reactive protocols, such as Ad hoc On-Demand Distance Vector (AODV) [12] and Dynamic Source Routing (DSR) [13], create routes on demand, which reduces overhead but can increase latency in large networks. While these methods provide basic routing solutions, they often lack the adaptability needed for complex and rapidly changing network environments.

In traffic prediction, traditional techniques such as statistical models and time series analysis methods (e.g., Autoregressive (AR), Moving Average (MA), ARMA, and ARIMA, [9]) have been widely used. These models are effective for basic predictions in relatively stable environments but struggle to capture the complexities of modern, dynamic networks. The non- linear and chaotic traffic patterns in contemporary networks render these methods less effective.

The primary drawback of these traditional methods is their inability to handle the non-linear, chaotic, and highly variable nature of traffic in modern networks, driven by event-based applications and real-time services.

5 Centralized ML Solutions

In response to these challenges, AI techniques, such as ML, are gaining popularity in network management. ML techniques can capture non-linear relationships and dynamically adjust to variations. In this section, we review recent ML-based solutions for congestion control, routing management, and traffic prediction.

5.1 Congestion Control

Several researchers have explored Machine Learning (ML) approaches for optimizing congestion control, with a strong focus on reinforcement learning (RL) techniques. One notable trend is the integration of RL with traditional networking protocols. For example, Abbasloo et al. [14] proposed a hybrid model combining the traditional TCP (BBR) protocol with an RL agent to dynamically adjust the congestion window. While this approach demonstrates potential for enhancing network responsiveness, it operates within a centralized framework, which raises concerns about scalability as network sizes increase.

Similar challenges arise in the work Midhula and Kumar [15], where an RL-based method for wireless networks was developed to improve throughput. Although the results indicate performance improvements, the need for real-world validation highlights the gap between experimental outcomes and practical deployment. This underscores a common limitation across many RL- based approaches, where theoretical gains need to be tested in more varied and complex network environments.

Beyond RL, other approaches have sought to leverage alternative ML techniques in specific contexts. For instance, Goteti et al. [16] and Andrade-Zambrano et al. [17] explored the use of Q-learning and Deep Q-Networks (DQN) in Wireless Sensor Networks (WSNs) and smart grids. These methods offer dynamic path selection and congestion control, showing potential to enhance network efficiency. However, as with RL, the lack of large-scale testing raises questions about their scalability and robustness in real-world conditions, particularly in more heterogeneous network architectures.

Meanwhile, Elbery et al. [24] introduced Machine Learning Aided Congestion Control (MLACC), a hybrid approach that blends heuristic-based adjustments with ML to predict bandwidth availability. This technique successfully reduces latency, but its reliance on predefined heuristic parameters limits its flexibility. Such parameters may struggle to adapt to the varied and unpredictable conditions characteristic of modern networks, a recurring challenge in ML-based congestion control.

In a different approach, Ahamed et al. [25] employed Genetic Algorithms (GA) to manage congestion in Vehicular Ad Hoc Networks (VANETs) by adjusting transmission power and data rates. Although the algorithm proved effective in simulations, scalability concerns remain, particularly in mitigating channel congestion in larger, more dense networks. Table 2 provides a summary of the advantages and limitations of these approaches.

In conclusion, while RL, DRL, and other heuristic-driven ML approaches have led to significant advancements in congestion control, a consistent set of challenges emerges across these methods: ensuring scalability, achieving real-world validation, and maintaining adaptability in the face of dynamic and diverse network conditions. These techniques, while innovative and promising, still require further refinement and empirical testing before they can be fully realized in large-scale, operational environments.

Table 2. Centralized ML Solutions for Congestion Control

Reference	Network Environment	ML model	Advantage	Limitation
[14]	General TCP	RL + traditional TCP (BBR)	Adaptive congestion window adjustment in varied scenarios	Centralized learning limits scalability in unseen environments
[25]	VANETs	Genetic algorithms	Reduces latency, optimizes transmission power and data rates	Faces channel congestion challenges in large networks
[16]	WSNs	Q-learning	Dynamically avoids congestion based on residual energy	Requires large-scale testing to confirm effectiveness
[17]	Smart Grids	DQN	Effective real-time state information for congestion control	Needs different neural network architectures for optimization
[24]	Inter- datacenter Networks	Heuristic + ML	Reduces latency by predicting bandwidth availability	Relies on predefined heuristic parameters, limiting adaptability
[26]	SDNs, Data Centers	Soft Actor Critic (SAC)	Real-time adaptation of congestion window	Requires further training in heterogeneous environments
[27]	Data Center Power Networks	DPG + POMDP	Outperforms traditional algorithms in fairness and utilization	Only tested in simulations, needs real hardware integration

5.2 Routing Management

Machine learning (ML) has become a powerful tool for routing tasks in communication networks, with diverse approaches designed to tackle dynamic traffic patterns, scalability, and real-time adaptability. Despite varying methodologies, they share a goal of enhancing routing efficiency, network performance, and quality of service (QoS).

In path management, Chung et al. [33] introduced MPTCP-ML, a random forest-based scheme that dynamically optimizes Wi-Fi path quality. While it shows clear improvements in signal strength and throughput, its scalability and robustness in real-world environments remain to be tested. This underscores the challenge of translating promising models from controlled experiments to large-scale deployments.

In Software-Defined Networking (SDN), Graph Neural Networks (GNN) have emerged as a sophisticated approach to improving routing decisions. Zhu et al. [34] proposed GCLR, which uses GNNs to optimize Multipath TCP (MPTCP) routing by predicting throughput based on network topology. However, prediction accuracy remains a limitation. Similarly, Ipek et al. [22] used various ML models, with Logistic Regression outperforming others for dynamic route selection. While these works demonstrate ML's role in SDN, challenges like prediction accuracy and real-world validation persist.

Reinforcement Learning (RL) and Deep RL (DRL) have also advanced routing optimization. Yajadda et al. [38] combined shortest path principles with RL to adjust routes based on real-time traffic intensity. While effective, its handling of multicast traffic and congestion control requires improvement. DRL has also been applied to SDN, as seen in Yu et al.'s [42].

DROM model, which continuously updates routing decisions. However, maintaining QoS in large-scale deployments remains challenging. In wireless mesh networks (WMNs), RL models like Sun et al.'s [44] optimize Packet Delivery Probability (PDP), but frequent updates and complexity may impact scalability.

RL's flexibility is evident in specialized networks like the Internet of Medical Things (IoMT), where Nazari et al. [47] developed an RL-based routing mechanism for medical applications, though real-world testing is still required.

Hybrid models, like Almasan et al.'s [18] integration of GNN with DRL agents, show improvements in routing efficiency, though adding more network state features could enhance performance. Similarly, their extension to IP networks [20] balances computational cost and optimization gains. In quantum communication networks, Roik et al. [23] used Proximal Policy Optimization (PPO) to outperform classical algorithms, though its performance against other ML techniques remains uncertain.

Table 3 summarizes the key advantages and limitations of these approaches.

Table 3. Centralized ML solutions for Routing

Reference	Network Environment	ML model	Advantage	Limitation
[33]	MPTCP, Wi-Fi	Random Forest-Based Path Management	Improves signal strength, data rate, and throughput	Requires real-world validation
[34]	SDN, MPTCP	GNN for Routing Optimization	Predicts throughput, adapts to varying topologies	Faces challenges in prediction accuracy

(continued)

Table 3. (*continued*)

Reference	Network Environment	ML model	Advantage	Limitation
[22]	SDN	AI for Dynamic Route Selection	Logistic Regression improves dynamic route selection	Needs broader real-world validation
[38]	General Networks	RL + Shortest Path Routing	Dynamically adjusts routes based on traffic intensity	Limited exploration of multicast and uniform traffic patterns
[42]	SDN	DDPG-based Routing Optimization	Optimizes routing through continuous state analysis	Requires strategies for maintaining QoS
[44]	WMNs	RL for Packet Delivery Optimization	Improves PDP while adhering to end-to-end delay constraints	Frequent updates may limit scalability
[47]	Internet of Medical Things (IoMT)	RL-based QoS Routing	Adapts routing mechanism to meet QoS demands	Scalability and resource efficiency need further testing
[18]	General Networks	GNN + DRL for Routing o ptimization	Enhances routing with GNN models and DRL agents	Lacks additional network state features for better accuracy
[23]	Quantum Communication Networks	PPO	Optimizes routing in quantum error-prone networks	Lacks comparison with other ML techniques

These studies highlight the growing sophistication of ML-driven routing solutions, though challenges like scalability, prediction accuracy, and balancing computational overhead persist.

5.3 Traffic Prediction

Machine learning (ML) has greatly enhanced network management by enabling accurate traffic prediction, allowing for proactive decisions that improve network efficiency and performance.

In cellular networks, the AML-CTP framework [28] clusters traffic using density-based algorithms like KDC and DBSCAN. Models such as Linear Regression, SVR, Decision Tree, and LightGBM are applied to these clusters, reducing computational time and improving model selection. LTE-A data shows that this framework maintains accuracy during peak hours, making it highly scalable. Future work will focus on traffic steering techniques to further improve Quality of Service (QoS).

In optical fiber communication systems, [29] introduces an RNN and LSTM-based traffic prediction algorithm. This approach differentiates between actual traffic fluctuations and random variability, providing real-time data that helps prevent congestion and optimize resource allocation. The model significantly reduces mean absolute error compared to existing methods.

For 6G networks, [63] proposes a parallel CNN architecture for traffic prediction. Using the Matrix Format Method (MFM) to convert traffic data into image representations, this model enhances resource allocation and energy management. While promising, further research is needed to adapt it to the evolving 6G environment.

In fog-IoT architectures, [32] integrates edge caching and load balancing with ML-based traffic prediction and resource management. RNNs and LSTMs are used for forecasting, while Random Forests handle real-time load balancing. Future enhancements could involve hybrid methods, such as combining LSTMs with SVMs or incorporating reinforcement learning.

For 5G networks, [58] presents a hybrid traffic prediction model that combines Holt smoothing and time series transformers to manage network traffic in dynamic slices. By using digital twins for real-time traffic prediction, this model improves resource management, though further testing is needed for scalability in IIoT environments.

Table 4. Centralized ML solutions for Traffic Prediction

Reference	Network Environment	ML model	Advantage	Limitation
[28]	Cellular Networks	AML-CTP with clustering and ML models	Reduces training time, scalable for peak hour predictions	Further exploration of traffic steering for QoS is needed
[29]	Optical Fiber Net-works	RNNs, LSTMs for traffic prediction	Differentiates traffic patterns, lowers mean absolute error	Requires validation in real-time congestion prevention
[63]	6G Networks	CNN with Matrix Format Method (MFM)	Tailored to 6G traffic, improves resource allocation	Needs adaptation for evolving 6G environments
[32]	Fog-IoT Architectures	RNNs, LSTMs for traffic prediction, Random Forest for load balancing	Reduces latency, efficient resource distribution	Potential for hybrid methods with SVMs and RL for further optimization

(*continued*)

Table 4. *(continued)*

Reference	Network Environment	ML model	Advantage	Limitation
[58]	5G Networks	Holt Smoothing and Time Series Transformer	Improves resource allocation, dynamic 5G slice management	Effectiveness varies by traffic distribution, requires real-world testing

Table 4 summarizes the advantages and limitations of these centralized AI approaches for network management. These works highlight the diversity of ML-based traffic prediction across cellular, optical fiber, 6G, and fog-IoT networks. Each solution addresses specific challenges in its network environment, though real-world validation is needed to ensure efficiency and adaptability.

6 Distributed AI Solutions

While centralized ML solutions have shown promise in improving network performance, their reliance on centralized models limits scalability and adaptability, particularly in large and dynamic environments. To address these challenges, distributed AI approaches have emerged as powerful alternatives, leveraging decentralized intelligence to optimize traffic flow and enhance network performance in real time.

6.1 Congestion Control

Distributed AI approaches for congestion control have made significant strides, especially with the application of MARL and distributed ML models. In [66], MARL is used for network congestion management, where agents collaborate across network layers to adjust parameters such as queue lengths and sending rates. This approach demonstrates the potential of MARL in enhancing network performance under dynamic conditions, though its validation has been limited to simplified simulated environments, necessitating further research into its real-world applicability and scalability.

A similar application of MARL is seen in [59], where the authors propose a framework for managing congestion in software-defined networks (SDNs), focusing on both host- to-host and switch-controller congestion. In this setup, two agents independently optimize paths and load balancing based on local observations like bandwidth and jitter. The approach uses a cooperative Q-learning algorithm to adapt to changing network conditions. However, the authors acknowledge that performance comparisons with existing algorithms are still needed.

Another study, [68], presents a scalable DRL-based online routing algorithm (DRL-OR-S) within an SDN framework. Each node optimizes routing decisions independently based on factors like latency, throughput, and packet loss. Despite its scalability, the reliance on hop-by- hop routing may not be as resource-efficient as multi-path solutions.

Likewise, [69] extends the DDPG algorithm to MA-DDPG for decentralized congestion control in network switches, where each switch makes independent decisions to optimize traffic flow. This approach addresses millisecond-level traffic fluctuations but faces similar challenges as [70], which applies MARL to MPTCP networks for congestion control in IoT and 5G environments. Both methods introduce significant computational complexity and rely on a global state, limiting their effectiveness when network conditions or node behavior are not fully observable.

For green energy harvesting in massive machine-type communications (mMTC), [67] introduces DDRQN for congestion control by optimizing uplink power control. Devices adjust policies based on historical data to mitigate traffic congestion, while sharing a centrally learned Q-network. This decentralized approach reduces communication and computation costs, though the initial training overhead presents a challenge in resource constrained environments.

In high-speed data center networks (DCNs), the PET approach [64] leverages the Independent Proximal Policy Optimization (IPPO) algorithm to dynamically adjust ECN thresholds based on factors like queue length, data rate, and traffic composition. Operating within a decentralized training and execution framework, PET improves flow completion time and reduces queue length variance. However, further research is needed to evaluate PET's scalability in real-world deployments.

Table 5 summarizes the key advantages and limitations of these distributed AI approaches for congestion control.

Table 5. Distributed AI Solutions for Congestion Control

Reference	Network Environm ent	ML Model	Advantage	Limitation
[66]	General network	MARL	Collaborates across network layers to dynamically manage traffic and optimize network performance	Tested only in simplified simulations; needs real- world validation for scalability
[59]	SDNs	Cooperative Q-learning	Optimizes paths and load balancing, adapts to changing network conditions	Lacks comparison with existing Q-learning algorithms, limiting evaluation
[68]	SDNs	DRL based on hop-by- hop routing	Scalable DRL-based routing algorithm, improves routing efficiency based on latency, throughput, and packet loss	Hop-by-hop routing is less efficient compared to multi-path routing

(*continued*)

Table 5. (*continued*)

Reference	Network Environm ent	ML Model	Advantage	Limitation
[69]	Network switches	MA-DDPG	Each switch makes independent decisions to optimize traffic flow in real- time	High computational complexity and reliance on a global state
[70]	IoT	MARL	Manages subflows based on parameters like RTT and throughput, improving congestion control in dynamic environments	Significant computational complexity; reliance on a global state reduces effectiveness
[67]	Green energy harvesting mMTC	Deep DDQ-Learning	Optimizes uplink power control, reduces communication and computation costs	High initial training overhead in resource-constrained environments
[64]	High- speed (DCNs)	Independent PPO	Dynamically adjusts ECN thresholds, improving flow completion time and queue length variance	Requires further research to assess scalability in real-world deployments

Across these works, the use of distributed AI, particularly MARL, demonstrates significant potential for enhancing congestion control in diverse network environments. However, common challenges such as reliance on a global state, computational complexity, and the need for real world validation persist. These methods highlight the growing sophistication of AI-driven congestion control solutions while emphasizing the need for further exploration to ensure scalability, efficiency, and adaptability in practical applications.

6.2 Routing Management

The exploration of distributed AI approaches for network management has led to significant advancements, especially with the integration of RL and federated learning (FL) models. These methods are not only improving network efficiency and resource allocation but also reshaping real-time decision-making in complex network environments.

In WSNs, a standout method involves decentralized Q-learning, as proposed in [71]. Here, sensor nodes operate autonomously, optimizing energy efficiency and data aggregation through local interactions. This distributed decision-making model is particularly promising for scaling WSNs, as it reduces the need for central control. The potential for extending this model into larger, more complex environments is clear, but real-world testing is crucial to validate its effectiveness in large-scale deployments.

Federated learning (FL) also plays a transformative role, especially in UAV networks. In [72], the B.A.T.M.A.N. protocol integrates FL, enabling UAVs to collaboratively learn from localized data. This method reduces bandwidth consumption and ensures data privacy while making routing decisions more adaptable to dynamic topologies. The promise of this decentralized approach lies in its ability to optimize routing decisions without overwhelming network resources, though its efficacy in more complex, real-world UAV networks remains to be proven. A similar approach is seen in [73], where FL is used to enhance resource allocation in UAV-based edge computing for 5G networks. The role of FL in managing next-generation networks is further highlighted by the QoSPR protocol in [78], which applies federated RL to 5G-IIoT networks. By decentralizing the task of routing and resource management to regional centers, this method preserves privacy and improves network performance. Challenges such as balancing latency and load across a larger scale still require further investigation, particularly as these systems scale to larger deployments in industrial applications. While these FL-based approaches demonstrate great potential, the models need further refinement to handle the complexities of real-world environments.

Another key component of distributed AI, reinforcement learning, is playing an increasingly pivotal role in advancing network management strategies. For instance, in [74], the distributed RL model (DCRL-R) allows WSN nodes to make autonomous routing decisions based on local data, thus minimizing reliance on a global network view. This is especially valuable in infrastructureless environments, though the model's scalability in larger networks needs validation. Similarly, in maritime networks, [75] applies multi-agent deep Q-learning to improve routing efficiency. The collaborative learning between agents provides a more informed decision- making process, but further testing in variable environments is needed to ensure robustness. Federated learning continues to show its versatility in network management. In [76], CNN-based agents are employed to monitor and predict congestion in WMNs. These agents analyze the network state, allowing for proactive rerouting and congestion avoidance. While this approach has demonstrated its potential in simulations, its effectiveness in real-world scenarios with more diverse network conditions requires further exploration. In [79], multi-agent deep RL is utilized for packet routing in computer networks, where each router acts independently based on local and neighbouring data. This decentralized method optimizes packet delivery efficiency, yet the simulations fail to capture the full complexity of real-world traffic, indicating a need for more comprehensive testing to ensure practical applicability.

Lastly, in the context of IoT networks, [77] tackles energy efficiency using MARL. IoT devices autonomously make routing decisions, focusing on sustainable network operations. While this approach shows great promise for optimizing energy use in IoT systems, the computational overhead of deploying MARL at scale remains a considerable challenge.

Table 6 summarizes the key advantages and limitations of these distributed AI approaches for routing.

As distributed AI techniques continue to mature, they offer the potential to transform traditional network management strategies. These innovative approaches promise to enhance the scalability, adaptability, and efficiency of communication networks, but further real-world testing and optimization are essential to unlock their full capabilities. The journey towards smarter, more autonomous networks is just beginning, and these advancements lay the foundation for a more intelligent and efficient future in network management.

Table 6. Distributed AI Solutions for Routing

Reference	Network Environment	ML Model	Advantage	Limitation
[71]	WSNs	Decentralized Q-Learning	Autonomously optimizes energy efficiency and data aggregation	Requires empirical validation for large-scale networks
[72]	UAV Networks	FL with B.A.T.M.A.N protocol	Maintains data privacy, reduces bandwidth, and adapts to changing topologies	Needs real-world validation in complex UAV scenarios
[73]	5G UAV-based Edge Computing	FL, RL	Efficient data routing and resource allocation	Simplified models, needs further testing under realistic conditions
[74]	Wireless Sensor Networks (WSNs)	Distributed	Decentralized decision- making, ideal for infrastructureless environments	Scalability in large networks unproven
[75]	Maritime Networks	Multi-Agent Deep Q-Learning	Improves routing efficiency via agent collaboration	Needs further testing in highly variable environments
[76]	Wireless Mesh Net works	CNN	Predicts congestion and dynamically reroutes traffic	Requires broader real-world testing
[78]	5G-IIoT Networks	FL	Preserves data privacy, optimizes performance for regional challenges	Needs real-world validation in large-scale applications

(*continued*)

Table 6. *(continued)*

Reference	Network Environment	ML Model	Advantage	Limitation
[79]	Computer Networks	MA-DRL	Independent routing decisions based on local data	Simulations do not fully capture real-world network complexities
[77]	IoT Energy-Harvesting Networks	MARL	Promotes energy-efficient routing and sustainable IoT development	High computational overhead in large-scale IoT networks

6.3 Traffic Prediction

Shifting focus to traffic prediction, distributed AI techniques are becoming instrumental in forecasting network traffic patterns and optimizing resource management, ensuring more efficient and proactive network operations.

[60] tackles the challenge through an innovative use of transfer learning. Instead of relying on traditional centralized models that require extensive data collection, this distributed approach trains models at the Network Operations Center (NOC) before fine-tuning them locally at each VSAT. This method effectively reduces bandwidth consumption while maintaining high accuracy, demonstrating how distributed AI can efficiently balance resource constraints and prediction precision in bandwidth-limited environments.

Table 7. Distributed AI Solutions for Traffic Prediction

Reference	Network Environment	ML model	Advantage	Limitation
[60]	Satellite Communication Networks	Transfer Learning	Reduces bandwidth usage while maintaining high prediction accuracy	Limited exploration of real-world scalability and generalization across diverse satellite conditions
[61]	Internet Traffic Networks	Distributed Multi-Agent Lear ning (LSTM/GRU)	Enhances prediction performance through decentralized, cooperative agent interaction	Fixed weight matrices limit adaptability in more complex scenarios (e.g., grid or graph-format predictions)

(continued)

Table 7. *(continued)*

Reference	Network Environment	ML model	Advantage	Limitation
[62]	Cellular Networks	Federated Learning (FL) with Sustainability Indicator	Balances energy consumption with prediction accuracy, minimizing environmental impact	Further research needed on model convergence speed and scaling FL across more diverse datasets and clients
[63]	at Network Edges	Federated Learning (LSTM)	Improves privacy, reduces processing time and bandwidth usage in decentralized setups	Handling non-IID data across clients remains a challenge, requiring utility optimization for fairness
[65]	Multi-Agent Network Automation System (MANA-NMS)	LSTM for Traffic Prediction, Decision Trees for Classification	Autonomous agents improve prediction and classification accuracy, aiding resource management	Needs further integration of additional network services such as routing and QoS monitoring for a complete system

A fully decentralized framework is explored by [61], which introduces a multi-agent learning system for Internet traffic prediction. Each agent operates autonomously, training its own LSTM or GRU models, with predictions synchronized through a collaborative Markov chain process. This decentralized structure not only enhances prediction performance but also ensures scalability across agents. However, the reliance on fixed weight matrices limits flexibility when handling complex, dynamic network topologies, highlighting a potential area for future enhancements in distributed AI traffic prediction systems.

[62] takes a different approach with federated learning (FL) in cellular traffic prediction, addressing concerns around privacy and data transfer limitations. The introduction of a sustainability indicator highlights the growing need to balance the environmental cost of training AI models with accuracy, especially in large-scale distributed systems. While FL allows for distributed training without sharing raw data, its environmental footprint, particularly for larger models, poses a significant challenge, prompting future research into optimizing the balance between energy efficiency and model performance.

For 5G networks, [63] further leverages the advantages of FL for LSTM-based traffic prediction at the network edge. This decentralized setup reduces processing time and enhances privacy compared to traditional centralized methods. However, the challenge of managing non- IID data across clients remains, impacting prediction fairness and consistency. Future efforts focused on utility optimization functions could address this limitation, ensuring a more equitable distribution of resources across clients while preserving accuracy in distributed AI systems.

Lastly, [65] proposes a multi-agent network automation architecture (MANA-NMS) that combines LSTM for traffic prediction with Decision Trees for traffic classification. These autonomous agents improve resource allocation and load balancing in network management. However, the system's full potential has yet to be realized, with further work needed to integrate aadditional network services, such as routing and QoS monitoring, into the framework, thus leveraging the power of distributed AI for more comprehensive network management.

The Table 7 summarizes the advantages and limitations of distributed AI solutions for traffic prediction, highlighting each approach's contribution to network management as well as areas that require further development.

In conclusion, distributed AI approaches for traffic prediction show great promise in enabling smarter and more adaptive network management across a variety of communication environments. These techniques not only improve efficiency and accuracy but also offer solutions to privacy, scalability, and environmental challenges. As the field continues to evolve, refining these systems and testing them in real-world conditions will be key to fully realizing their potential in optimizing network performance.

7 Conclusions

In this paper, we have explored the evolving landscape of AI-driven solutions for complex communication networks, covering both centralized and distributed approaches in congestion control, routing management, and traffic prediction. While centralized solutions perform well in controlled environments, distributed AI techniques stand out for their scalability, adaptability, and potential to operate effectively in dynamic, large-scale network environments. However, despite their promise, distributed AI approaches also face challenges; more work is needed to refine these techniques and address these limitations.

Future research will continue to optimize these AI-driven approaches to meet the growing demands of modern communication networks. To tackle these ongoing challenges, our current research focuses on developing and evaluating a Multi-Agent Deep Reinforcement Learning (MA-DRL) traffic management solution. This approach is designed to dynamically adapt to network conditions, prevent congestion, and improve key metrics such as throughput, latency, and packet loss, providing a more robust and adaptive alternative to existing traffic management methods while addressing some of the current limitations of distributed AI.

References

1. Boccaletti, S., Latora, V., Moreno, Y., Chavez, M., Hwang, D.-U.: Complex networks: structure and dynamics. Phys. Rep. **424**(4), 175–308 (2006). https://doi.org/10.1016/j.physrep.2005.10.009
2. Dorogovtsev, S.N., Goltsev, A.V., Mendes, J.F.F.: Critical phenomena in complex networks. arXiv:0705.0010 [cond-mat.stat-mech] (2007). Last revised Nov. 2007, https://doi.org/10.48550/arXiv.0705.0010

3. Albert, R., Barabasi, A.-L.: Statistical mechanics of complex networks. arXiv:cond-mat/0106096 [condmat.stat-mech]. Submitted to Reviews of Modern Physics **74**, 47 (2002). https://doi.org/10.48550/arXiv.condmat/0106096
4. Newman, M.E.J.: The structure and function of complex networks. SIAM Rev. **45**, 167–256 (2003). https://doi.org/10.1137/S003614450342480
5. Eddy, W.M.: Transmission Control Protocol (TCP). RFC **9293**, 1–98 (2021)
6. Dong, X., Stewart, R.R., Tu¨xen, M.: ECN for Stream Control Transmission Protocol (SCTP). Proceedings of the IEEE (2014)
7. Mandala, S., Iskandar, M.N., Nugroho, M.A.: Active Queue Management (AQM) Performance Analysis Based on Controlled Delay (CoDel) Algorithm on User Datagram Protocol (UDP). In: Proceedings of the International Conference on Information Technology Systems and Innovation (ICITSI) (2017)
8. Patel, S., Shukla, Y.D., Kumar, N., Sharma, T., Singh, K.: A Comparative Performance Analysis of TCP Congestion Control Algorithms: NewReno, Westwood, Veno, BIC, and Cubic. In: 2020 6th International Conference on Signal Processing and Communication (ICSC), pp. 23–28 (2020)
9. ArunKumar, K.E., Kalaga, D.V., Mohan, Ch., Kumar, S., Kawaji, M., Brenza, T.M.: Comparative analysis of Gated Recurrent Units (GRU), Long Short-Term Memory (LSTM) cells, Autoregressive Integrated Moving Average (ARIMA), Seasonal Autoregressive Integrated Moving Average (SARIMA) for forecasting COVID-19 trends. Alex. Eng. J. **61**, 7585–7603 (2022)
10. Perkins, C.E., Bhagwat, P.: Highly dynamic Destination-Sequenced Distance-Vector routing (DSDV) for mobile computers. In: Proceedings of the Conference on Communications Architectures, Protocols and Applications (1994)
11. Open Shortest Path First (OSPF). Enterprise Certification Study Guide (2020)
12. Majumder, A., Kumar, A., Fdida, S.: Ad-hoc On-demand Distance Vector. In: Proceedings of the International Conference (2013)
13. Johnson, D.B., Maltz, D.A., Broch, J.: DSR: The dynamic source routing protocol for multihop wireless ad hoc networks. In: Proceedings of the International Conference (2001)
14. Abbasloo, S., Yen, C.-Y., Chao, H.J.: Classic meets modern: a pragmatic learning-based congestion control for the internet. In: Proceedings of the Annual Conference of the ACM Special Interest Group on Data Communication on the Applications, Technologies, Architectures, and Protocols for Computer Communication (2020)
15. Midhula, M.K.S., Arun Raj Kumar, A.R.K.P.: An adaptive congestion control protocol for wireless networks using deep reinforcement learning. IEEE Transactions on Network and Service Management **21**, 2027–2043 (2024)
16. Goteti, D., Rasheed, I.: An approach to integrate reinforcement learning in wireless sensor networks to evade congestion. In: 2023 International Conference on Sustainable Communication Networks and Application (ICSCNA), pp. 58–65 (2023)
17. Andrade-Zambrano, A.R., et al.: A reinforcement learning congestion control algorithm for smart grid networks. IEEE Access **12**, 75072–75092 (2024)
18. Almasan, P., Sua´rez-Varela, J., Rusek, K., Barlet-Ros, P., Cabellos-Aparicio, A.: Deep reinforcement learning meets graph neural networks: Exploring a routing optimization use case. Computer Communications **196**, 184–194 (2022)
19. Ferriol-Galmés, M., et al.: RouteNet-Fermi: Network modeling with graph neural networks. IEEE/ACM Transactions on Networking (2023)
20. Almasan, P., Sua´rez-Varela, J., Rusek, K., Barlet-Ros, P., Cabellos-Aparicio, A.: A multi-agent reinforcement learning perspective on distributed traffic engineering. In: 2021 IEEE 28th International Conference on Network Protocols (ICNP) (2021)
21. Jiang, W., He, M., Gu, W.: Internet traffic prediction with distributed multi-agent learning. Applied System Innovation **5**(6), Art. no. 121 (2022)

22. Ipek, A.D., Cicioglu, M., Calhan, A.: Enhancing video transmission with machine learning-based routing in software-defined networks. In: 2024 IEEE Conference on Communications (ICC) (2024)
23. Roik, J., Bartkiewicz, K., Cernoch, A., Lemr, K.: Routing in quantum communication networks using reinforcement machine learning. Quantum Inf. Process. **23**, 89 (2024)
24. Elbery, A., Lian, Y., Li, G.: Toward fair and efficient congestion control: Machine learning aided congestion control (MLACC). In: Proceedings of the 7th Asia-Pacific Workshop on Networking (2023)
25. Ahamed, V.M.N., Sivaraman, K., Bharath, Ph.D.: Congestion control system optimization with the use of vehicle edge computing in VANET powered by machine learning. International Journal of Computer Networks and Applications (2024)
26. Galliera, R., Morelli, A., Fronteddu, R., Suri, N.: MARLIN: Soft Actor-Critic based reinforcement learning for congestion control in real networks. In: NOMS 2023–2023 IEEE/IFIP Network Operations and Management Symposium, pp. 1–10 (2023)
27. Huang, T., Lu, X., Zhang, D., Cheng, H., Dong, P., Zhang, L.: ACC-RL: Adaptive congestion control based on reinforcement learning in power distribution networks with data centers. Energies (2023)
28. Nashaat, H., Mohammed, N.H., Abdel-Mageid, S.M., Rizk, R.Y.: Machine learning-based cellular traffic prediction using data reduction techniques. IEEE Access **12**, 58927–58939 (2024)
29. Malhotra, A.: A new traffic prediction algorithm with machine learning for optical fiber communication system. In: 2023 3rd Asian Conference on Innovation in Technology (ASIANCON), pp. 1–7 (2023)
30. Tamada, K., Kawamoto, Y., Kato, N.: Bandwidth usage reduction by traffic prediction using transfer learning in satellite communication systems. IEEE Transactions on Vehicular Technology **73**, 7459–7463 (2024)
31. Bolivar, R.P.M., Kumar, S.K.N., Priya, V., et al.: 6G traffic prediction with a novel parallel convolutional neural networks architecture and matrix format method integration. J. Machine and Computing (2024)
32. Anu, D., S., R., Mageshwari, P.U.: Optimizing fog-IoT networks: A synergistic approach using LSTM traffic prediction and random forest load balancing. Int. J. Res. Publ. Rev. (2024)
33. Chung, J., Han, D., Kim, J., Kim, C.-K.: Machine learning-based path management for mobile devices over MPTCP. In: 2017 IEEE International Conference on Big Data and Smart Computing (BigComp), pp. 206–209 (2017)
34. Zhu, T., Chen, X., Chen, L., Wang, W., Wei, G.: GCLR: GNN-based cross-layer optimization for multipath TCP by routing. IEEE Access **8**, 17060–17070 (2020)
35. Yang, L., Wei, Y., Yu, F., Han, Z.: Joint routing and scheduling optimization in time-sensitive networks using graph-convolutional-network-based deep reinforcement learning. IEEE Internet Things J. **9**, 23981–23994 (2022)
36. Xu, Z., Tang, J., Yin, C., Wang, Y., Xue, G.: Experience-driven congestion control: When multi-path TCP meets deep reinforcement learning. IEEE J. Sel. Areas Commun. **37**, 1325–1336 (2019)
37. Maller, R.A., Müller, G., Szimayer, A.: Ornstein–Uhlenbeck processes and extensions. In: Proceedings of the Conference (2009)
38. Yajadda, S.H., Safaei, F.: A novel reinforcement learning routing algorithm for congestion control in complex networks. In: Proceedings of the Conference (2023)
39. Kong, Y., Zang, H., Ma, X.: Improving TCP congestion control with machine intelligence. In: Proceedings of the 2018 Workshop on Network Meets AI ML (2018)
40. Kanagarathinam, M.R., Singh, S., Sandeep, I., Roy, A., Saxena, N.; D-TCP: Dynamic TCP congestion control algorithm for next-generation mobile networks. In: *2018 15th IEEE Annual Consumer Communications Networking Conference (CCNC), pp. 1–6 (2018)

41. Nie, X., et al.: Dynamic TCP initial windows and congestion control schemes through reinforcement learning. IEEE J. Selec. Areas in Comm. **37**, 1231–1247 (2019)
42. Yu, C., Lan, J., Guo, Z., Hu, Y.: DROM: Optimizing the routing in software-defined networks with deep reinforcement learning. IEEE Access **6**, 64533–64539 (2018)
43. Rischke, J., Sossalla, P., Salah, H., Fitzek, F.H.P., Reisslein, M.: QR-SDN: Towards reinforcement learning states, actions, and rewards for direct flow routing in software-defined networks. IEEE Access **8**, 174773–174791 (2020)
44. Sun, W., Wang, Z., Zhang, G.: A QoS-guaranteed intelligent routing mechanism in software-defined networks. Comput. Networks **185**, 107709 (2021)
45. Duong, T.V.T., Binh, L.H., Ngo, V.M.: Reinforcement learning for QoS-guaranteed intelligent routing in wireless mesh networks with heavy traffic load. ICT Express **8**, 18–24 (2022)
46. Yun, W.-K., Yoo, S.-J.: Q-learning-based data-aggregation-aware energy-efficient routing protocol for wireless sensor networks. IEEE Access **9**, 10737–10750 (2021)
47. Nazari, A., Kordabadi, M., Mohammadi, R., Lal, C.: EQRSRL: An energy-aware and QoS-based routing schema using reinforcement learning in IoMT. Wireless Networks **29**, 3239–3253 (2023)
48. Xu, Z., et al.: Experience-driven networking: A deep reinforcement learning based approach. In: IEEE INFOCOM 2018 - IEEE Conference on Computer Communications, pp. 1871–1879 (2018)
49. Sun, P., Hu, Y., Lan, J., Tian, L., Chen, M.: TIDE: Time-relevant deep reinforcement learning for routing optimization. Future Gener. Comput. Syst **99**, 401–409 (2019)
50. Fang, J., et al.: Reinforcement learning for bandwidth estimation and congestion control in real-time communications. ArXiv, vol. abs/1912.02222 (2019)
51. Astudillo Leon, J.P., de la Cruz Llopis, L.J., Rico-Novella, F.: A machine learning based distributed congestion control protocol for multi-hop wireless networks. Comput. Networks **231**, 109813 (2023)
52. Tian, H., Liao, X., Zeng, C., Zhang, J., Chen, K.: Spine: An efficient DRL-based congestion control with ultra- low overhead. In: Proceedings of the 18th International Conference on emerging Networking EXperiments and Technologies (2022)
53. Bai, J., Zhang, T., Xie, G.: MACC: Cross-layer multi-agent congestion control with deep reinforcement learning. ArXiv, vol. abs/2206.01972 (2022)
54. He, B., et al.: DeepCC: Multi-agent deep reinforcement learning congestion control for multi-path TCP based on self-attention. IEEE Trans. Netw. Ser. Manage. **18**, 4770–4788 (2021)
55. Han, X., Han, B., Li, R., Ji, X.: MARS: An adaptive multi-agent DRL-based scheduler for multipath QUIC in dynamic networks. In: 2023 IEEE/ACM 31st International Symposium on Quality of Service (IWQoS), pp. 01–10 (2023)
56. Cash, M., Ali, H.: WIP-FL: A federated learning-based protocol for UAV networks. In: 2023 IEEE International Conference on Communications (ICC) (2023)
57. Watanabe, Y., Kawamoto, Y., Kato, N.: A novel routing control method using federated learning in large-scale wireless mesh networks. IEEE Trans. Wireless Comm. **22**, 9291–9300 (2023)
58. Min, Z., et al.: A novel 5G digital twin approach for traffic prediction and elastic network slice management. In: 2024 16th International Conference on COMmunication Systems & NETworkS (COMSNETS), pp. 497–505 (2024)
59. Boussaoud, K., Ayache, M., En-Nouaary, A.: A Multi-Agent Reinforcement Learning Approach for Congestion Control in network based-SDN. In: 2023 10th International Conference on Future Internet of Things and Cloud (FiCloud), pp. 250–255 (2023)
60. Tamada, K., Kawamoto, Y., Kato, N.: Bandwidth Usage Reduction by Traffic Prediction Using Transfer Learning in Satellite Communication Systems. IEEE Trans. Vehic. Technol. **73**, 7459–7463 (2024)

61. Jiang, W., He, M., Gu, W.: Internet Traffic Prediction with Distributed Multi-Agent Learning. Applied System Innovation **5** (2022)
62. Perifanis, V., Pavlidis, N., Yilmaz, S.F., et al.: Towards Energy-Aware Federated Traffic Prediction for Cellular Networks. In: 2023 Eighth International Conference on Fog and Mobile Edge Computing (FMEC), pp. 93–100 (2023)
63. Bolivar, R.P.M., Kumar, S.K.N., Priya, V., et al.: 6G Traffic Prediction with a Novel Parallel Convolutional Neural Networks Architecture and Matrix Format Method Integration. J. Machi. Comp. (2024)
64. Cheng, K., Wang, T., Du, X., Du, S., Cai, H.: PET: Multi-agent Independent PPO-based Automatic ECN Tuning for High-Speed Data Center Networks. ArXiv, vol. abs/2405.11956 (2024)
65. Arzo, S.T., Akhavan, Z., Esmaeili, M., et al.: Multi-Agent-Based Traffic Prediction and Traffic Classification for Autonomic Network Management Systems for Future Networks. Future Internet **14**(230) (2022)
66. Bai, L., Zhang, Y., Wang, X.: MA-CC: Multi-Agent Congestion Control Mechanism for Wireless Networks Using Reinforcement Learning. IEEE Communications Letters **26**(7), 1345–1349 (2022)
67. Teng, J., Liu, Y., Zhang, H.: Distributed Deep Reinforcement Learning for Uplink Power Control in Green Energy-Harvesting mMTC. IEEE Trans. Comm. **68**(12), 7500–7512 (2020)
68. He, Y., Tan, J., Zhang, Z.: A Scalable DRL-based Online Routing Algorithm for Software-Defined Networks. IEEE Access **11**, 146789–146801 (2023)
69. Mai, X., Zhang, H., Li, F.: Adaptive DDPG for Distributed Congestion Control in Network Switches. IEEE Trans. Netw. Ser. Manage. **17**(2), 1143–1155 (2020)
70. He, Z., Zhu, H., Zhang, J.: DeepCC: multi-agent reinforcement learning for congestion control in dynamic IoT networks. IEEE Internet of Things J. **8**(10), 8556–8568 (2021)
71. Yun, H., Kim, B.: Q-Learning-Based Decentralized Energy-Efficient Routing for Wireless Sensor Networks. IEEE Access **9**, 7429–7439 (2021)
72. Cash, M., Ali, H.: WIP-FL: A Federated Learning Based Protocol for UAV Networks. In: 2023 IEEE International Conference on Communications (ICC) (2023)
73. Singh, N., Patel, P.: 5G and Beyond: Federated Learning for UAV Edge Computing. IEEE Trans. Wireless Comm. **21**(4), 2323–2331 (2023)
74. Barker, D., Jones, A.: Distributed reinforcement learning for routing optimization in wireless sensor networks. IEEE Sensors J. **22**(5), 3187–3198 (2022)
75. Yang, L., Zhang, T., Sun, K.: Multi-Agent Deep Q-Learning for Routing in Maritime Networks. IEEE Internet of Things J. **10**(1), 234–245 (2023)
76. Watanabe, T., Sato, H., Matsuda, K.: A novel routing framework for wireless mesh networks using federated learning with convolutional neural networks. IEEE Access **11**, 6703–6716 (2023)
77. Zhang, Z., Yu, F., Zhang, R.: Energy Harvesting-Aware Routing for IoT Networks Using Multi-Agent Reinforcement Learning. IEEE Internet of Things J. **9**(13), 10435–10446 (2022)
78. Wang, J., Tao, L., Han, S.: QoSPR: Federated Reinforcement Learning for Gateway Deployment and Routing in Industrial IoT Networks. IEEE Trans. Indus. Info. **18**(7), 5097–5108 (2022)
79. You, K., Li, J.: Toward packet routing optimization in distributed networks using multi-agent deep reinforcement learning. IEEE Trans. Netw. Ser. Manage. **19**(2), 230–243 (2022)

A Framework for Global Trust and Reputation Management in 6G Networks

Position Paper

Bac Trinh-Nguyen[1,2(✉)], Sara Berri[1], Sin G. Teo[3], Tram Truong-Huu[4], and Arsenia Chorti[1]

[1] ETIS UMR 8051/CY Paris University, ENSEA, CNRS, Paris, France
{trinh-nguyen.bac,sara.berri,arsenia.chorti}@ensea.fr
[2] CNRS, IPAL, Singapore, Singapore
[3] Agency for Science, Technology and Research (A*STAR), Singapore, Singapore
teosg@i2r.a-star.edu.sg
[4] Singapore Institute of Technology (SIT), Singapore, Singapore
truonghuu.tram@singaporetech.edu.sg

Abstract. The evolution of the sixth generation (6G) of wireless will enable the full-scale deployment of systems of cyber-physical systems (CPS) such as autonomous vehicles, robots, and drones. Currently, a paradigm shift in security and trust-building approaches for 6G is being discussed within the communications and networking communities, to tackle novel and more sophisticated behavioral attacks of malicious CPS agents. As an example, such autonomous agents will use vast amounts of data – generated from and exchanged amongst heterogeneous network devices – for artificial intelligence-driven decision-making. As a result, standard device authentication approaches and current agent reputation models will not suffice for trust building; the verification, traceability, and trustworthiness of the exchanged data themselves become critical issues. In this paper, we highlight related open issues using as an example a specific behavioral attack in a vehicular ad-hoc network (VANET), to open up the broader discussion regarding potential trust issues in 6G CPS networks. In light of this, we propose a comprehensive framework for trust and reputation management in 6G. We discuss alternative framework architectures and provide high-level details of the framework components with involved technologies.

Keywords: Cyber-physical systems · 6G · trust and trustworthiness · bogus information attacks · machine learning · trust and reputation management

1 Introduction

The rapid development of communication technologies in recent years paves the way for the evolution of 6G networks, which are expected to contribute to

F. Hacène et al. (Eds.): MLN 2024, LNCS 15540, pp. 55–72, 2026.
https://doi.org/10.1007/978-3-032-00552-6_4

the interconnected world of the future. These future networks will be deployed in various application domains such as smart cities, smart healthcare, industrial automation, and more, forming cyber-physical systems (CPS) where agents seamlessly interact with both the physical and digital worlds and are also known as *multi-agent cyber-physical systems* [16]. In such a dynamic ecosystem, autonomous agents, vehicles, robots, and other intelligent devices will be able to communicate and make decisions independently of humans. However, the increased connectivity and interaction between autonomous entities will open up novel potential security risks and challenges [30,34].

One key challenge stems from the fact that these agents will exchange data continuously, including regarding their *physical states* such as location, speed, acceleration, and environmental conditions, to assist in making accurate real-time decisions. Multiple challenges and opportunities arise as a result. For example, apart from the obvious location privacy issues, a key challenge concerns how to verify that the exchanged information is accurate, not modified or intentionally falsified as it propagates through multiple agents. To capture related issues, the ITU 2018 report [26] defines trustworthy networking as *a set of methods to provide reliable and secure communications among any pair of network elements that have trust relationships.* Subsequent works [22] identified six 6 attributes for trustworthy systems, namely *resilience, security, privacy, safety, reliability, and availability.* In addition, a vast amount of literature is dedicated to trustworthy AI [21], while high-level views of the requirements for 6G trust were also presented [32]. Without robust trust-building mechanisms, malicious agents can conduct adversarial attacks, inject false information that leads to privacy concerns, or cause potentially widespread corruption across the network.

In this paper, we take the position that trust and reputation models should *use both views of the physical world* – as provided through sensing and the lower communication layers – *and the cyber world*, as provided through semantics and upper communication layers, unlike existing trust and reputation models. To highlight the importance of incorporating physical views in trust models, in a preliminary investigation, we exemplify security attacks in VANETs, a typical example of CPS where network operations take place in a highly dynamic environment. The mobility of vehicles and constant changes in network topology highlight the need for trust and reputation management to maintain network security. One of the common threats in VANETs is the bogus information attack [14] or a false data injection attack [9], where rogue vehicles send false data with the intention of causing delays, traffic congestion, accidents, or gaining priority. As a proof of concept, we discuss the practicality and effectiveness of a multi-view trust framework incorporating location and speed information.

Motivated by the previous discussion, we highlight the potential open issues in 6G trust and trustworthiness and propose a novel framework for global trust and reputation management (TRM). We present two potential architectural approaches that take into account cross-layer information from the physical layer to the application layer, as well as sensing and semantic plane inputs.

The rest of the paper is organized as follows. In Sect. 2, we present our preliminary investigation with VANETs, to motivate the follow-up discussion. In Sect. 3, we discuss the new trends in 6G and describe the paradigm shift in trust building and management to meet the evolving requirements of 6G networks of CPSs. In Sect. 4, we present possible trust management framework architectures with detailed descriptions of the framework components. In Sect. 5 we conclude the paper and present future directions.

2 Experiments and Motivation

Along with the development of 6G networks, the need to expand VANETs is also constantly increasing, leading to security risks. Attacks in VANETs focus on different directions to undermine confidentiality, integrity, availability, identity, authentication, and non-repudiation properties [33]. Figure 1 shows the network stack layers including attack vectors classified as single-layer or multi-layer attacks, along with the unique characteristics of each layer used for current detection solutions. At each layer, we identify the features that can be abused by adversaries to perform attacks but also be used by defenders to detect attacks or abnormal behavior.

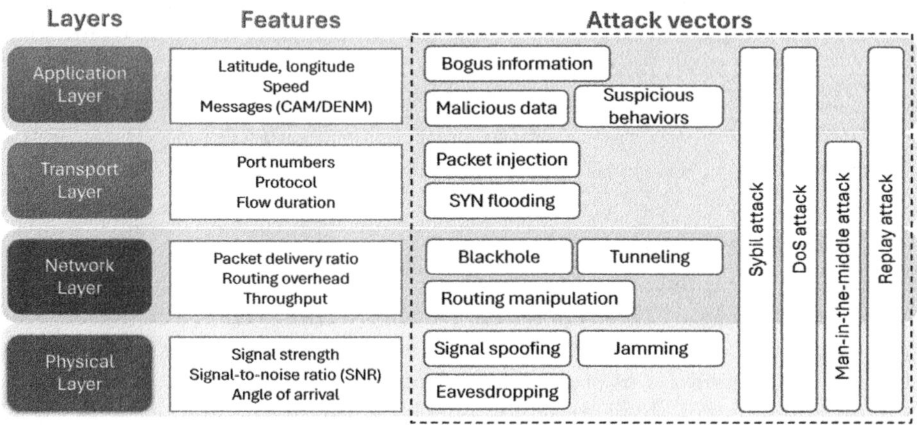

Fig. 1. Features and attack vectors in each layer of the network stack.

- **Physical layer.** In the study by Alzahrani *et al.* [4], received signal strength indicator (RSSI) and angle of arrival (AoA) are used as trusted features to determine the reliability of cooperative awareness messages (CAM) received from adjacent vehicles. However, we note that the RSSI and the AoA have been shown to be vulnerable to certain types of attacks, e.g., when using analog array processing chains [29] as opposed to digital array processing [24].

- **Network layer.** Shamim *et al.* [35] used features such as packet delivery ratio (PDR), routing overhead (ROH), throughput, packet loss rate (PLR), packets generated, packets dropped, energy consumption in their collaborative detection method of black hole and gray hole attacks.
- **Transport layer.** Hassan *et al.* [17] used the features at the transport layer to interpret the decision-making process of the random forest classifier, thereby understanding the characteristics of different types of attacks in VANETs. For example, "destination-port" represents the port number where the packet will arrive, "flow-packets-sec" is the packet rate in a flow, and "min-seg-size-forward" is the minimum segment size in the forward direction. However, it has been shown that destination ports could be vulnerable to a certain type of attacks, such as web-attack, infiltration, and port scan (PS) [23].
- **Application layer.** So *et al.* [28] calculated velocity, acceleration, and distance traveled values from GPS location information and then converted them into plausible checks to detect anomalies. Several attacks on GPS location have been identified, along with several mitigation methods [10,31]

To highlight the novel, complex security challenges in 6G networks of CPSs, in this section, we illustrate a specific use case of anomaly detection in a VANET. First, we use the Eclipse MOSAIC simulation platform [27] to set up different VANET scenarios. Then, we examine the limitations of current approaches and present a proposed solution using positioning and speed information to identify a bogus information attack. As an example, an attacker attempts to broadcast a fake message about obstacles on the current route using a type of vehicle-to-everything (V2X) communication called decentralized environmental notification message (`DENM`) to disperse vehicles away from the optimal route to gain an advantage [19].

2.1 Scenario Setup

Simulation Framework. We used Eclipse MOSAIC [27] to develop a small-scale attack scenario. This simulation framework allows for the realistic simulation of vehicle communications, providing a controlled environment and entities for testing different types of attacks and mitigation approaches. It also enables the programming of applications for vehicles and roadside units (RSUs).

Simulation Setting. The experimental scenario takes place in an urban setting around the area of the ETIS campus in Cergy, France, with a simple road topology, including a main route and an alternative route. One of the vehicles in the middle of a convoy plays the role of an attacker, broadcasting fake `DENM` messages to notify about non-existing obstacles on the main route, causing vehicles within a certain radius to redirect to an alternative (sub-optimal) route. During the simulation, the system collects real-time data exchanged between vehicles and infrastructure. The collected data includes information on location, speed, and acceleration, and all communication is monitored for analysis.

Attack Mitigation. We implemented a centralized system with a misbehavior detection mechanism, specifically focusing on detecting bogus information attacks within this scenario. The simulation system includes:

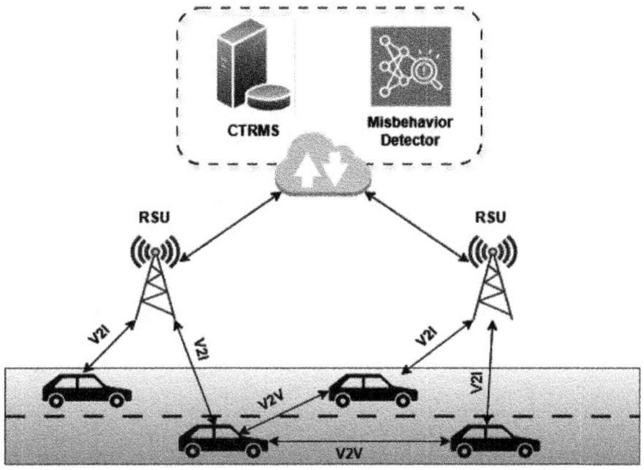

Fig. 2. Illustration of simulation scenario including V2X communications: vehicle-to-vehicle (V2V) and vehicle-to-infrastructure (V2I).

- **A centralized trust and reputation management system (CTRMS)**: The system collects logs from the network and then transfers them to the central server for storage and processing. The system includes a rule-based anomaly detector and vehicle trust management, which is quantified as a trustworthiness score to describe the truthfulness of the vehicle in the network. The management model is shown in Fig. 2.

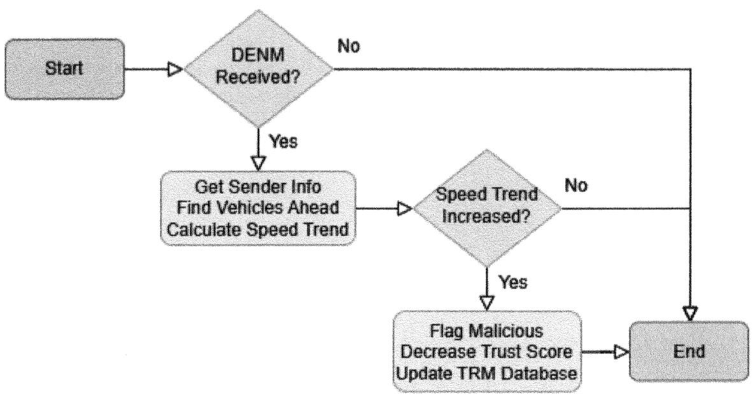

Fig. 3. Flowchart for rule-based detection of bogus information attack.

- **Misbehavior detector**: This module plays a role in analyzing the collected data of vehicles in the network and finding anomalies, identifying malicious vehicles. In the scope of this scenario, we implemented a simple rule-based detection mechanism. This rule-based processing flow is described in Fig. 3. When a vehicle receives a `DENM` message from the network, it requests CTRMS to check the sender's trustworthiness. The system then verifies whether there are any abnormalities based on the collected logs and identifies the location of the vehicles and the behavior (speed and movement trends) of nearby vehicles. If any abnormalities related to the sender are detected, the system will flag the sender as malicious, decrease their trust score, notify the requesting vehicle, and take necessary measures. In this scenario, to simplify, all vehicles start with a trust score of 1. Once a vehicle is identified as malicious, its trust score will decrease to 0. Based on this, vehicles receiving the `DENM` message will decide whether to change routes according to the trust scores provided by the CTRMS system.

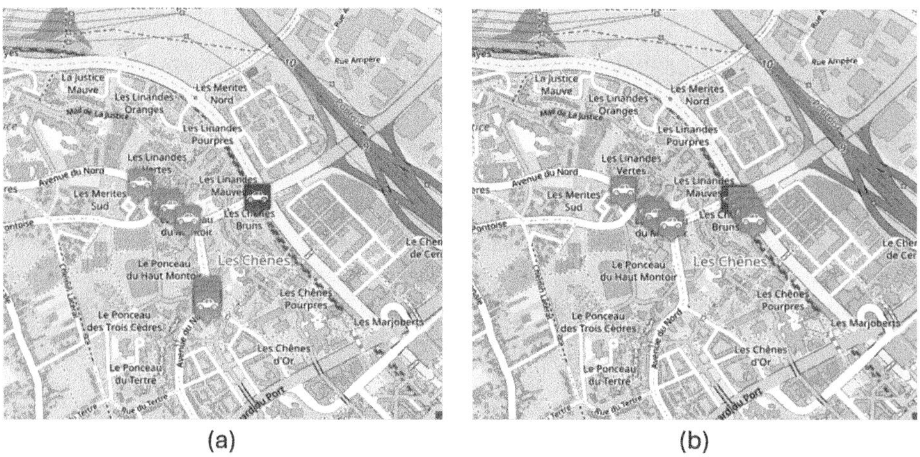

Fig. 4. Map visualization for two test cases **(a) successful bogus information attack successfully** and **(b) unsuccessful attack – detected and prevented**. The red marker indicates the malicious vehicle while the green markers represent benign vehicles.

2.2 Analysis

We conducted experiments through three test cases: the first simulates an attack without CTRMS, the second demonstrates the detection capability of the CTRMS combined with a misbehavior detector, and the third highlights the scalability issue of the system when there is a large number of vehicles in the network. We measure the processing delay for each request at CTRMS.

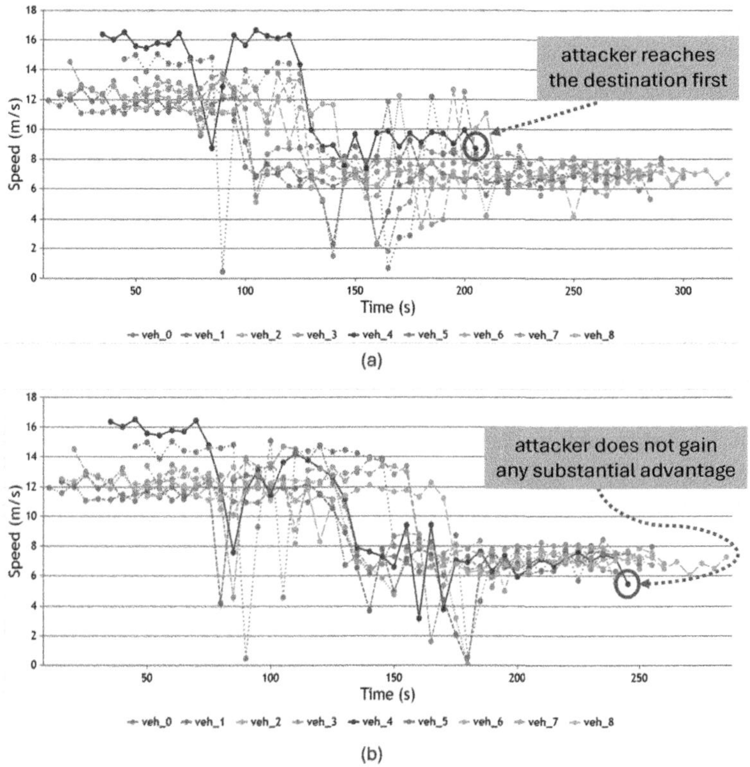

Fig. 5. Vehicle speed graph for two test cases **(a) bogus information attack successfully propagated** and **(b) attack detected and prevented.**

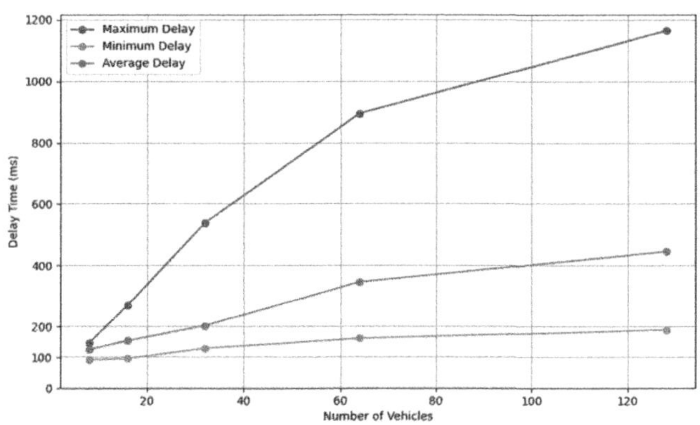

Fig. 6. Delay time measurement (maximum, minimum, average) for different vehicle counts (8, 16, 32, 64, 128).

- **Case 1: Successful attack.** Due to the lack of a trust framework, the attacker can succeed in spreading bogus information about the optimal route to the neighboring vehicles, causing them to change to the alternative route, shown in Fig. 4(a). In this toy example, the attacker's gains correspond to reaching the destination first, at around 200 s after departure, while the other vehicles arrive within the time frame of 300 s. The impact on vehicle speed is illustrated in Fig. 5(a).
- **Case 2: The attack is detected and mitigated.** With the CTRM system deployed along with a misbehavior detector, the attack was successfully detected. Through the trust score verification from the server, the decision-making units on other vehicles will not make the rerouting decision, and the attack was successfully mitigated, as shown in Fig. 4(b). As a result, the attacker did not gain any substantial advantage, the speed graph is displayed in Fig. 5(b).
- **Case 3: Delay time measurement.** We continue with the above scenario but increase the number of vehicles in the network from 8 to 16, 32, 64, and 128, respectively. The purpose of this task is to simulate varying traffic densities in the network and, consequently, collect and analyze the processing time and response time (delay time) from the centralized server to the requested vehicles based on these density variations. The results are shown in Fig. 6, including the maximum, minimum, and average delay times. We notice that when the number of vehicles in the network increases, the server processing and response time slows down significantly. This exceeds the maximum requirement for low latency services, typically requiring from 1ms to 10ms [13,20]. The low latency or ultra-low latency is an important element in 6G applications such as self-driving cars, remote surgery, and industry automation. In the context of VANETs, failing to meet the strictly end-to-end latency requirements between vehicles, roadside units (RSU), and the traffic control center (TCC) can lead to serious consequences. VANETs are designed to support real-time communication for safety-critical applications such as collision avoidance, lane change assistance, and accident prevention.

Through this above example, we can observe that implementing a centralized system to manage the trustworthiness scores of nodes in the network shows both strengths and limitations in detecting anomalous behavior using a rule-based approach. *Firstly*, the system's dependence on application-layer data such as GPS, speed, and acceleration information can reduce its effectiveness in identifying complex attack patterns. If GPS spoofing is successful, the attack cannot be identified simply by this type of misbehavior detection mechanism. *Secondly*, using static rules to identify attacks is limited due to the difficulty of handling dynamic and evolving attack strategies, e.g., involving the coordinated action of multiple malicious entities simultaneously (colluding adversaries). *Thirdly*, as the size of the network increases, the rule-based system becomes less efficient in processing real-time data, leading to detection delays and an increased likelihood of false positives and false negatives. Thus, although similar approaches have been proposed in the literature [30], the limited trust and reputation model used proves to be inefficient for 6G networks of CPSs, where a vast number of autonomous nodes will connect, communicate, and actuate, using large volumes of complex data.

In such environments, the system must be able to adapt in real-time and handle multi-layer and multi-modality data to identify threats accurately. ML/DL models offer promising solutions for processing large amounts of data, identifying complex attack patterns and anomalies, and enabling the detection of both known and unknown (zero-day) attacks. Moreover, by combining the data from multiple layers, sensing and semantic layers, ML/DL models can comprehensively analyze the network and the node behaviors, providing the system with a vital view to detect the increasingly diverse and evolving threats in 6G networks.

3 Security Challenges and Opportunities in 6G

Below, we enumerate several key challenges that will impact the design of security solutions in 6G, notably with respect to latency, quantum resistance, and the role of AI. Subsequently, we outline novel opportunities that will be brought up in 6G, in terms of multiple security levels (quality of security), the introduction of physical layer solutions, and integrated communications and sensing.

3.1 New Security Challenges in 6G Networks

Low Latency, Low Footprint, Scalable Security: Ensuring robust security for networks of autonomous agents requires operating within strict latency limits, handling massive connectivity, and maintaining low energy consumption, low footprint and low computational load, is an ambitious challenge.

Quantum Resistance: Additionally, to remain secure against future threats, systems must incorporate quantum-resistant cryptographic methods. This approach aligns with the newly standardized post-quantum cryptographic algorithms established by NIST [5,6], the adoption of quantum-resistant algorithms into IoT devices is challenging due to their limited resources.

AI and ML: The widespread deployment of AI and ML in 6G systems will expand potential vulnerabilities, highlighting the need for defenses against adversarial AI (such as data-poisoning attacks). Additionally, the energy footprint needs to be contained to sustainable levels (green AI) and the AI outputs need to be interpretable, and unbiased (XAI) [8].

3.2 Opportunities for Strengthened Security in 6G

Alongside these challenges, new security opportunities emerge, including decentralized and democratized approaches to learning and computation, such as federated learning, the integration of sensing capabilities, and the ability to perform context and semantics distillation. These advancements open doors to innovative technologies, like physical layer security, as discussed below.

The Introduction of Sensing in 6G: Positioning and radar sensing will be default features in 6G networks [2]. This integration creates possibilities for using sensing and positioning for integrity checks and anomaly detection that takes into consideration the physical behavior devices. While Sybil cyberattacks in robotic systems have been found using AoA [15], a number of techniques that use location information as a second soft authentication factor have already been developed. Trusted positioning will be a vital aspect of assessing the trustworthiness of autonomous agents in 6G. However, this also introduces privacy challenges.

Quality of Security (QoSec): In the long run, a safe and sustainable future requires adaptability to maximize limited resources. In this context, adaptive security protocols and algorithms could be developed to adjust configurations and parameters dynamically, based on multi-layered inputs from sensing, semantic, and contextual data. QoSec offers a scalable security framework for CPS networks, providing varying levels of security and trust to meet the needs of future networks.

Physical Layer Security: Within an adaptive, AI-driven security framework, PLS solutions [12], which leverage physical phenomena for security, are envisioned as complements to post-quantum cryptography, enhancing 6G's trustworthiness and resilience. PLS could enable secure keyless communication or the generation and distribution of symmetric keys through wireless channel properties. We will be able to systematically make use of such opportunities in 6G thanks to channel engineering and controllability (e.g., with the use of metasurfaces, drones for multi-hop networks, and very narrow beamforming). Additionally, PHY and hardware layer authentication, e.g., utilizing physical unclonable functions and localization as a second-factor authentication, can allow for rapid, continuous verification of legitimate users, minimizing the need for upper-layer processing.

3.3 Paradigm Shift in Security for 6G Networks

The development of the new generation of networks has promoted the transition from a network of connected things to an intelligent network, where agents have more autonomous capabilities and more interactions with each other. However, without TRM systems in multi-agent networks (MAS) or IoT, such interactions will not ensure trustworthiness and will pose many security risks [36]. Therefore, 6G networks need to be designed with inherent trustworthiness to improve security effectiveness and enhance privacy protection capabilities [18].

Trust and Trustworthiness: The behavior of devices, agents, and systems must be considered holistically when establishing trust and assessing trustworthiness. For a trustworthy 6G, multiple layers of trust must be guaranteed. While ongoing discussions continue, it is evident that a secure 6G network must ensure trust in both the AI intelligence and the infrastructure, including sensing, communication channels, and data processing. The initial anchors of trust can be reduced to trusting at a highly abstract level: (i) The sensing (radar, RF, camera, lidar, etc.) that gathers raw or processed data (high precision localization information is especially crucial in 6G) and drives actuation; (ii) The platforms for computation and processing (including learning and optimization) at various network nodes (on devices, edges, and core networks); (iii) The communication links that transport data and authenticate agents and devices, offering assurances of confidentiality, integrity, authentication, and availability; (iv) The AI algorithms that define the actions of autonomous agents, devices, and systems based on data inputs and sensor readings.

In summary, 6G will be the first generation inherently built on AI, connecting intelligent and autonomous cyber-physical systems, digital twins of physical entities, and the metaverse. This integration of the physical, digital, and human realms signifies a new era where the physical properties of interconnected systems are essential for security.

4 Trust and Reputation Management Framework

In 6G networks, AI is expected to be native, meaning it plays a vital role in the network infrastructure as a fundamental component of the operations, administration, and management process across all layers of the system [1]. A high-level perspective on the components and factors maintaining trustworthiness is presented in Fig. 7. The data plane, control plane, and sensing plane are accountable for data processing, network management, and environmental monitoring. The data plane handles forwarding and routing packets in the network, using AI to actively monitor data flows, identify inconsistencies, and automatically flag them, thus enhancing data security and integrity. The control plane manages routing decisions and overall network operations. AI enhances the control plane by offering real-time analytics and adaptive decision-making algorithms that continuously evaluate the trustworthiness of network entities such as routers, switches, or RSUs in VANETs. The sensing plane gathers data from the physical environment through sensors (e.g., GPS, speed, acceleration) and other contextual inputs. AI enhances trust at the sensing plane by validating sensor data in real-time, identifying anomalies, and quickly detecting faulty or compromised sensors. This reduces the risk of false information entering the network, improves sensor accuracy, and minimizes potential vulnerabilities. Besides, semantics is a key topic emerging in 6G with the semantic plane performing signal processing, and information filtering across all layers of the protocol stack by leveraging ML algorithms deployed on edge and core cloud platforms [25].

One of the critical challenges is how to quantify trustworthiness within a network. In this section, we present two possible approaches for measuring trustworthiness through the idea of the TRM framework. These frameworks play a crucial role in analyzing historical data, evaluating the behavior of network entities, and

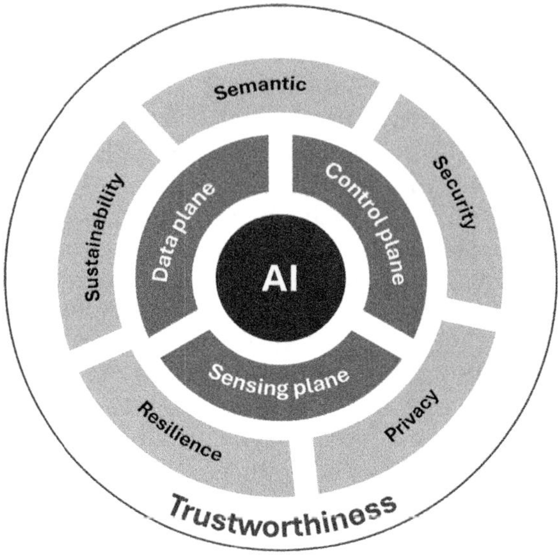

Fig. 7. High-level overview of trustworthiness in 6G.

real-time assessment through the integration of AI-based techniques. The first one is a multi-layer trust management system, as described in Sect. 4.1, where a global TRM serves as the highest authority, aggregating trust from all layers through their respective local TRM. The local trust evaluation is supported by ML/DL algorithms, which analyze features at each corresponding layer. On the other hand, the second one is a multimodal attention fusion approach, as outlined in Sect. 4.2, which leverages ML/DL to process and analyze data, extract the most important features at each layer, and synthesize them through a fusion module to support the process of assessing trustworthiness. This approach provides a comprehensive view of context and multiple behavior aspects, making trust assessments more reliable and accurate.

4.1 Multi-layer Trust and Reputation Management Framework

This framework is designed to maintain network security by combining multi-layer data from physical layer (PHY) signals and the information from the upper layers, applying ML and DL techniques to improve the capability of detecting attacks or misbehavior while enhancing the system's robustness. In Fig. 8, we present our first possible framework architecture, which is based on a multi-layer approach. The general framework architecture includes multiple TRM modules, each being responsible for collecting and processing data captured at each respective layer from PHY to higher network layers. Each layer plays an important role in the process of detecting potential attacks by providing unique features and observations at that layer to the inference module integrated with ML/DL models for analyzing complex attacks, handling vast amounts of data in real

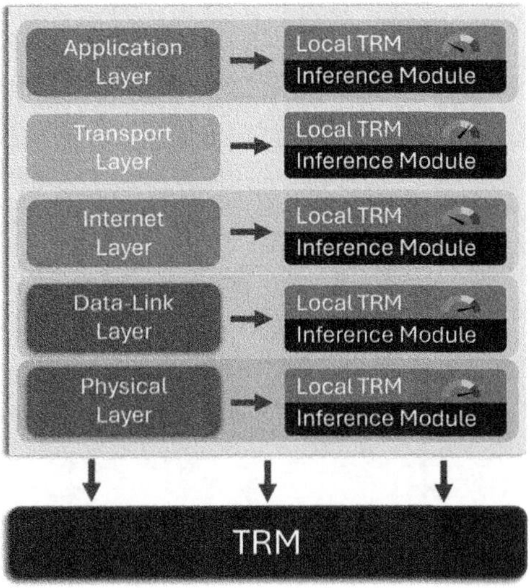

Fig. 8. Multi-layer architecture for TRM framework.

time, identifying subtle deviations in behavior that traditional rule-based systems often miss. At each layer, we propose to implement a specific local TRM based on the decision results from the inference module. Next, a global TRM is deployed to aggregate the decision from all local TRMs and compute an overall trust score for each node or vehicle. Depending on the nature of the attacks, a weightage vector can be applied to prioritize the decision of a particular local TRM. This multi-layer TRM architecture provides a holistic view and comprehensive trust score across the entire network while ensuring trust evaluation according to the data type and attack vector characteristics of each layer. The flexibility of this architecture is expected to be compatible with many types of cyber-physical systems while ensuring robustness in various contexts.

By combining the general model and the specific use case in the context of VANETs, Fig. 8 emphasizes the need for multi-layered security and trustworthiness in future autonomous and interconnected systems.

4.2 Multimodal Attention Fusion Approach

Because of the variety of data types of wireless communication characteristics, it is not appropriate to apply a specific ML/DL algorithm to calculate trust. We propose to develop and deploy additional planes including communication, sensing, and semantic planes to collect further information from the environment and correlate them with the data collected from the network. For example, the data collected in VANET can include GPS coordinates, velocity, acceleration, weather sensor data, camera images from the sensing plane; signal strength, packet transmission rates from the communication plane, and the meaning and context behind the data being transmitted from the semantic plane. Therefore,

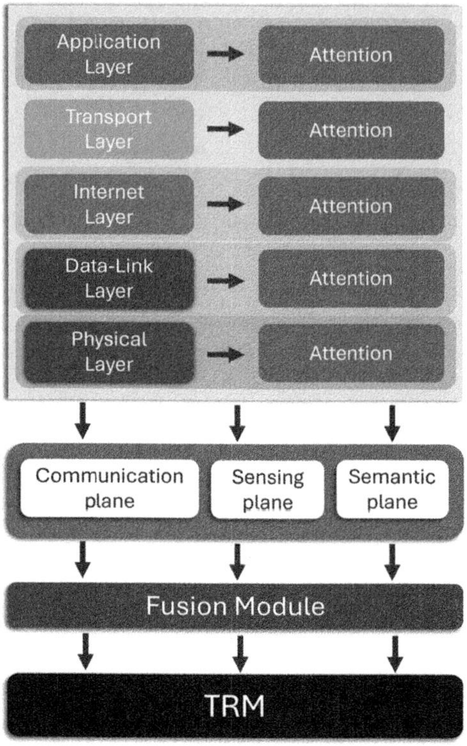

Fig. 9. Multimodal attention fusion architecture for TRM framework.

we propose to use a multimodal approach [7] which can use multiple types of input data for processing and analysis, providing a comprehensive view and contextual understanding, which has been applied and researched in security solutions [3,11]. Figure 9 depicts our proposed framework based on the idea of combining the attention mechanism and multimodal models. In this multimodal security framework, TRM can benefit from the attention models by focusing on trustworthiness data at each layer. Unlike the architecture presented in Sect. 4.1, the main idea of this approach is to utilize the attention mechanism to highlight key features at each layer, and then combine (fuse) the features into a unified representation based on the correlation among features. This allows the model to consider multiple perspectives simultaneously, increasing the accuracy of decision-making. The proposed framework includes the following components:

- **Data collection and attention to various layers:** The model applies attention to the data at each layer, identifying and prioritizing the most relevant information that is most important for a particular task. For example, with the task of detecting attacks such as jamming or spoofing at the PHY layer, the attention model could focus on specific features from a diverse set of data collected at the physical layer such as signal strength (RSSI), signal-

to-noise ratio (SNR), bit error rates (BER), channel state information (CSI). By applying the attention mechanism, the model can focus on specific signal patterns that indicate fake signal transmission or jamming attempts.
- **Fusion module (multimodal models):** Once the important features at each layer are weighted using the attention mechanism, all features will be aggregated into a unified representation based on their correlation. This allows the model to consider multiple perspectives and contexts, thus detecting complex, multi-vector attacks.
- **Final decision-making:** After synthesizing the attention-weighted features from multiple layers, the model makes predictions or other decisions. The trust scores of nodes in the network are calculated based on insights from information across multiple layers. This allows for comprehensive and accurate decision-making.

The proposed framework is expected to form an effective trust and reputation management solution and enhance the ability to detect sophisticated, multi-layered attacks by fusing attention-weighted features from different layers of the network stack. In addition, combining it with AI-based methods will help improve the system's resilience against both known and unknown attack vectors in 6G networks.

5 Conclusion

In this paper, we introduced a global trust and reputation management framework for 6G networks through two possible architectural approaches, a multi-layer approach and a multimodal attention fusion approach. The proposed multimodal attention fusion framework is expected to enhance security in multi-layer networks by integrating attention mechanisms with a global trust and reputation management system. It is especially suitable in the context of 6G networks where massive connectivity and ultra-low latency, high security, and reliability are required. The scalability of the multimodal fusion approach allows it to handle vast amounts of collected from numerous entities across multiple network layers. Integrating a global TRM system will further enhance decision-making by continuously adjusting security responses based on the behavior, trustworthiness, and past actions of network entities. Based on the ideas proposed in this study, some potential directions of TRM framework development include:

- Development of a practical TRM framework: Since implementing trustworthiness in technical systems is still in its early stages, the next phase will focus on building a real-world TRM framework, where ML/DL algorithms play a crucial role in handling and analyzing data across multiple network layers. Besides, it also needs to ensure fairness and interpretability in the decision-making processes of these algorithms, as transparency is essential for trust in autonomous systems. Additionally, the developed framework will prioritize improving detection accuracy for anomalies while minimizing false positives. This will involve refining the ML/DL models to better differentiate between

normal variances in behavior and actual malicious activities, ultimately leading to a more robust and dependable TRM framework.
- Define and develop trust scoring mechanisms for the entities and their connections within the network: These mechanisms should be flexible and able to dynamically and continuously update based on changes in the entities' behaviors, enabling a more adaptive and sustainable trust management system.
- Considering the practical applicability and scalability of the system, deploying the TRM framework in large-scale cyber-physical systems presents a significant challenge. It will need to ensure scalability, meet latency requirements, and operate within resource constraints while maintaining accurate anomaly detection capabilities. Therefore, it is essential to implement realistic scenarios that can provide insights into the framework's performance across diverse conditions and environments. These scenarios will enable fine-tuning of the model to suit the specific requirements of each system.

Acknowledgments. B. Trinh-Nguyen has been supported by the IPAL project CONNECTING. A. Chorti and S. Berri have been partially supported by the EC through the Horizon Europe/JU SNS project ROBUST-6G (Grant Agreement no. 101139068), the EU HORIZON MSCA-SE TRACE-V2X project (Grant No. 101131204), the ANR-PEPR 5G Future Networks projects, and the CYU INEX-PHEBE and NESLI projects.

References

1. Transforming the 6G vision to action. White paper, Nokia Bell Labs (2024)
2. 3GPP: study on integrated sensing and communication. Tech. Rep. TR 22.837, 3GPP (2022)
3. Agrafiotis, G., Kalafatidis, S., Giapantzis, K., Lalas, A., Votis, K.: Advancing cybersecurity with AI: a multimodal fusion approach for intrusion detection systems. In: 2024 IEEE International Mediterranean Conference on Communications and Networking (MeditCom) (2024)
4. Alzahrani, M., Idris, M.Y., Ghaleb, F.A., Budiarto, R.: Robust misbehavior detection scheme for vehicular network. In: 2021 International Conference on Data Science and Its Applications (ICoDSA) (2021)
5. Avanzi, R., et al.: CRYSTALS-Kyber algorithm specifications and supporting documentation. NIST PQC Round (2017)
6. Bai, S., et al.: CRYSTALS–Dilithium: algorithm specification and supporting documentation (2021)
7. Baltrušaitis, T., Ahuja, C., Morency, L.P.: Multimodal machine learning: a survey and taxonomy. IEEE Trans. Pattern Anal. Mach. Intell. **41**(2), 423–443 (2019)
8. Belle, V., Papantonis, I.: Principles and practice of explainable machine learning. Front. Big Data **4** (2021)
9. Bennet Praba, M.S., Rathna, R.: False data injection attack detection in VANET using upgraded grey wolf optimization algorithm using LSTM classifier. In: Sharma, H., Shrivastava, V., Bharti, K.K., Wang, L. (eds.) Commun. Intell. Syst., pp. 703–713. Springer Nature Singapore, Singapore (2023)

10. Burns, J., Amiridis, D., Kar, D.C., Li, L.: On effectiveness of machine and deep learning algorithms for detection of GPS spoofing attacks on unmanned aerial vehicles. In: 2023 Congress in Computer Science, Computer Engineering, & Applied Computing (CSCE), pp. 2405–2410. IEEE (2023)
11. Chebbi, S., Jebara, S.B.: Deception detection using multimodal fusion approaches. Multimedia Tools Appl. **82**(9) (2023)
12. Chorti, A., et al.: Context-aware security for 6G wireless: the role of physical layer security. IEEE Commun. Stand. Mag. **6**(1), 102–108 (2022)
13. European Telecommunications Standards Institute: ETSI TS 122 261 V15.9.0 (2021-10): Service requirements for next-generation networks; Stage 1. Tech. Rep. V15.9.0, ETSI (2021)
14. Ghaleb, F.A., Zainal, A., Maroof, M.A., Rassam, M.A., Saeed, F.: Detecting bogus information attack in vehicular ad hoc network: a context-aware approach. Procedia Comput. Sci. **163**, 180–189 (2019)
15. Gil, S., Kumar, S., Mazumder, M., Katabi, D., Rus, D.: Guaranteeing spoof-resilient multi-robot networks. Auton. Robot. **41**(6), 1383–1400 (2017). https://doi.org/10.1007/s10514-017-9621-5
16. Gil, S., Yemini, M., Chorti, A., Nedić, A., Poor, H.V., Goldsmith, A.J.: How physicality enables trust: a new era of trust-centered cyberphysical systems (2023)
17. Hassan, F., Yu, J., Syed, Z.S., Ahmed, N., Al Reshan, M.S., Shaikh, A.: Achieving model explainability for intrusion detection in VANETs with lime. PeerJ Comput. Sci. **9**, e1440 (2023)
18. Huang, X., et al.: 6G Trustworthiness Considerations. Tech. rep, NGMN Alliance (2023)
19. Iqbal, S., Ball, P., Kamarudin, M.H., Bradley, A.: Simulating malicious attacks on VANETs for connected and autonomous vehicle cybersecurity: a machine learning dataset. In: 2022 13th International Symposium on Communication Systems, Networks and Digital Signal Processing (CSNDSP), pp. 332–337. IEEE (2022)
20. Jun, S., Kang, Y., Kim, J., Kim, C.: Ultra-low-latency services in 5G systems: a perspective from 3G PP standards. ETRI J. **42**(5), 721–733 (2020)
21. Li, B., et al.: Trustworthy AI: from principles to practices. ACM Comput. Surv. **55**(9) (2023)
22. Li, J., Mao, B., Liang, Z., Zhang, Z., Lin, Q., Yao, X.: Trust and trustworthiness: what they are and how to achieve them. In: 2021 IEEE International Conference on Pervasive Computing and Communications Workshops and other Affiliated Events (PerCom Workshops), Kassel, Germany (2021)
23. Loumponias, K., Raptis, S., Darra, E., Tsikrika, T., Vrochidis, S., Kompatsiaris, I.: Forecasting cyber-attacks to destination ports using machine learning. In: ICISSP, pp. 757–764 (2023)
24. Pham, T.M., Senigagliesi, L., Baldi, M., Fettweis, G.P., Chorti, A.: Machine learning-based robust physical layer authentication using angle of arrival estimation. In: GLOBECOM 2023-2023 IEEE Global Communications Conference, pp. 13–18. IEEE (2023)
25. Popovski, P., Simeone, O.: Start making sense: semantic plane filtering and control for Post-5G connectivity. arXiv preprint: arXiv:1901.06337 (2019)
26. Recommendation ITU-T Y.3053: Framework of trustworthy networking with trust-centric network domains. Tech. Rep. ITU-T Y.3053, International Telecommunication Union (2018)
27. Schrab, K., et al.: Modeling an its management solution for mixed highway traffic with eclipse mosaic. IEEE Trans. Intell. Transp. Syst. **24**(6), 6575–6585 (2023)

28. So, S., Sharma, P., Petit, J.: Integrating plausibility checks and machine learning for misbehavior detection in VANET. In: 2018 17th IEEE International Conference on Machine Learning and Applications (ICMLA), pp. 564–571 (2018)
29. Srinivasan, M., Senigagliesi, L., Chen, H., Chorti, A., Baldi, M., Wymeersch, H.: AoA-based physical layer authentication in analog arrays under impersonation attacks. arXiv preprint: arXiv:2407.08282 (2024)
30. Sullivan, S., Brighente, A., Kumar, S.A.P., Conti, M.: 5G security challenges and solutions: a review by OSI layers. IEEE Access **9** (2021)
31. Wagner, M., Fröhlich, A.A.: Securing cyber-physical systems against GPS spoofing attacks using confidence attribution. In: 2023 International Conference on Software, Telecommunications and Computer Networks (SoftCOM), pp. 1–6. IEEE (2023)
32. Wang, Y., Kang, X., Li, T., Wang, H., Chu, C.K., Lei, Z.: SIX-Trust for 6G: toward a secure and trustworthy future network. IEEE Access **11** (2023)
33. Xu, X., Wang, Y., Wang, P.: Comprehensive review on misbehavior detection for vehicular ad hoc networks. J. Adv. Transp. **2022**(1) (2022)
34. Yan, W.K., Chan, R., Truong-Huu, T.: 5G core security: an insider threat vulnerability assessment. In: Proc. 13th Conference on Information Technology and Its Applications (CITA), Danang, Vietnam (2024)
35. Younas, S., Rehman, F., Maqsood, T., Mustafa, S., Akhunzada, A., Gani, A.: Collaborative detection of black hole and gray hole attacks for secure data communication in VANETs. Appl. Sci. **12**(23), 12448 (2022)
36. Zeynalvand, L., Luo, T., Zhang, J.: COBRA: context-aware Bernoulli neural networks for reputation assessment. In: Proceedings of the AAAI Conference on Artificial Intelligence **34**, 7317–7324 (2020)

DRL Framework for Minimizing Beam Switching Time and Maintaining QoS in 6G-V2X Base Stations

Ali Belgacem[1(✉)] and Abbas Bradai[2]

[1] Poitiers University, XLIM Research Institute, Poitiers, France
ali.belgacem@univ-poitiers.fr

[2] LEAT Lab, Cote d'Azur University, Nice, France
abbas.bradai@univ-cotedazur.fr

Abstract. This paper tackles the issue of reducing beamforming switching time while ensuring Quality of Service (QoS) in 6G Vehicle-to-Everything (V2X) communication systems. We introduce a Deep Reinforcement Learning (DRL) framework employing Proximal Policy Optimization (PPO) to optimize base station beamforming switching times, thereby improving communication efficiency and reliability for connected vehicles. By reducing switching time and maintaining QoS, our solution improves overall network performance, representing a notable advancement for autonomous and connected vehicle systems. Our findings demonstrate that PPO is a promising approach for enhancing V2X communication, optimizing network performance, and maintaining consistent QoS, setting the stage for future research in complex vehicular networks.

Keywords: 6G · Vehicle to Everything (V2X) · Beamforming · Proximal Policy Optimization · Base station · QoS · Beam switching time

1 Introduction

In 6G wireless communication systems, base stations play a critical role in managing signal transmissions, particularly in Vehicle-to-Everything (V2X) networks. Beamforming enhances signal quality by directing energy in specific directions, but the time required to switch between beamforming states can introduce delays, impacting both network performance and user experience-especially in fast-moving V2X environments. As 6G networks grow more complex and the demand for reliable V2X connectivity increases, optimizing beamforming transitions becomes essential. The challenge lies in managing these transitions efficiently while maintaining high-quality service, amid dynamic factors such as user mobility and fluctuating channel conditions in vehicular networks[8].

Deep Reinforcement Learning (DRL) has emerged as a powerful tool for addressing complex decision-making problems, particularly in communication

F. Hacène et al. (Eds.): MLN 2024, LNCS 15540, pp. 73–84, 2026.
https://doi.org/10.1007/978-3-032-00552-6_5

systems like Vehicle-to-Everything (V2X) networks. By integrating reinforcement learning with deep learning, DRL can effectively manage high-dimensional state spaces and optimize decisions to maximize long-term rewards. In V2X networks, DRL has been successfully applied to optimize resource allocation, power control, and interference management. Its adaptability to dynamic environments makes it especially suitable for minimizing beamforming switching time in base stations, which is crucial for maintaining reliable V2X communication [6].

Minimizing beamforming switching time focuses on optimizing transitions between predefined beamforming states to enhance network performance and the user experience in a V2X network. The primary goals are to reduce switching time, lowering latency, and improve system efficiency while maintaining or enhancing Quality of Service (QoS) for vehicles, measured by metrics like Signal-to-Interference-plus-Noise Ratio (SINR) or data rate. The key challenge is balancing these objectives, especially in dynamic network conditions where vehicle mobility and fluctuating channels complicate beamforming transitions [5].

This paper presents a DRL-based framework utilizing Proximal Policy Optimization (PPO) to optimize beamforming state switching in 6G base stations. It provides a comprehensive overview of the system model, objectives, and constraints, while also detailing the design of the DRL framework, including the definition of state-action spaces and reward functions. By addressing a critical challenge in 6G communication, this work offers a novel proposition that advances the application of DRL in wireless networks, aiming to enhance efficiency and performance in beamforming transitions.

To the best of our knowledge, no work in the literature has directly and explicitly addressed the challenge of optimizing beam switching time while simultaneously maintaining Quality of Service (QoS) in the context of V2X communication for 6G base stations. Existing studies either focus solely on QoS or on beam switching time, without considering both aspects together. For example, in [2], researchers investigated how integrating advanced antennas with Deep Reinforcement Learning (DRL) can enhance 6G vehicle-to-vehicle (V2V) communications. This study uses beam-switched antennas alongside DRL to optimize beam switching under dynamic conditions, aiming to improve throughput, delay, reliability, and spectral efficiency. On the other hand, in [7], the authors addressed optimizing V2X communication for 6G by enhancing QoS in subterahertz (100–300 GHz) communications. They tackle challenges such as high path loss and propose multi-hop relaying with different antenna placements. The windshield antenna location offers the best results, increasing coverage by 50% and reducing BS density by up to 32%. Although these methods improve upon traditional approaches, achieving real-time performance is still challenging. Optimizing beam switching time while maintaining QoS remains unresolved, as frequent handovers can increase switching times and degrade network performance. Further validation and research are needed to address these issues in V2X-6G base stations.

The structure of this paper is as follows: Sect. 2 introduces the problem formulation, outlining the system model and objectives. Section 3 details the

DRQL-based solution, covering the framework, state, and action spaces, reward function, and policy learning process. In Sect. 4, the steps to integrate the Proximal Policy Optimization (PPO) algorithm are outlined. Section 5 provides an overview of the simulation setup, performance metrics, and evaluation procedures. Lastly, Sect. 6 offers a conclusion.

2 Problem Statement and Formulation

Switching between beamforming states can introduce latency, which may degrade overall network performance. To address this issue, the goal is to minimize switching time while preserving or enhancing the QoS for vehicles. Optimizing these beamforming transitions with DRL is crucial for improving the efficiency and performance of vehicle communication systems. Therefore, we formulate this problem as follows:

Base Station (BS) Configuration

- n_t: The number of antennas at the base station (BS) used to form a beam.
- $\mathbf{B} = \{b_1, b_2, \ldots, b_m\}$: Set of beams. m represents the number of predefined beamforming states.
- $\mathbf{V} = \{v_1, v_2, \ldots, v_k\}$: Set of vehicles.

Beamforming State Switching

- $\tau(s_i, s_j)$: Time required to switch from beamforming state s_i to s_j, where $\mathbf{S} = \{s_1, s_2, \ldots, s_M\}$ represents the set of possible beamforming states.

Channel State Information (CSI)

- $\mathbf{H}_t$: Channel matrix at time t representing the channel gains between the BS and V.

Quality of Service (QoS)

Maintaining QoS requires ensuring that both the Signal-to-Interference-plus-Noise Ratio (SINR) and the data rate for each vehicle meet predefined thresholds to guarantee satisfactory communication performance.

- **Signal-to-Interference-plus-Noise Ratio (SINR)** It is a key metric in wireless communication that measures the quality of a received signal by accounting for interference and noise levels. For a vehicle V^k, the SINR is mathematically defined as:

$$\mathrm{SINR}_k = \frac{\varphi_p^k \times \varphi_G^k}{\varphi_I^k + \varphi_N^k}$$

where:

  - φ_p^k is the power of the signal received by vehicle k.
  - φ_G^k is the gain of the communication channel for vehicle k.
  - φ_I^k is the total interference power affecting vehicle k from other vehicles or sources.
  - φ_N^k is the noise power affecting vehicle k.
- **Data rate** (η) In beamforming, the data rate (η) is determined by the Shannon capacity formula based on the SINR at the receiver. It is given by:

$$\eta = W \log_2 (1 + \text{SINR})$$

  where:
  - W is the bandwidth of the channel

In order to ensure satisfactory service quality and performance, the following formulas define the minimum acceptable thresholds for SINR and data rate:

- `SINR Threshold`: The SINR for each vehicle k must satisfy:

$$\text{SINR}_k \geq \text{SINR}_{\text{min}}$$

  where SINR_{min} is the minimum acceptable SINR threshold.
- `Data Rate Threshold`: Similarly, the data rate for each vehicle k must satisfy:

$$\eta_k \geq \eta_{\text{min}}$$

  where $\eta_k$min is the minimum acceptable data rate threshold.

Therefore, QoS can be calculated by combining SINR and Data Rate using a formula:

$$\text{QoS} = \alpha \cdot \text{SINR} + \beta \cdot \eta$$

where:

- α and β are weights that determine the influence of SINR and Data Rate on the QoS score.

3 Deep Reinforcement Learning Framework

To achieve the stated objective, we propose using Deep Reinforcement Learning (DRL), which excels in complex decision-making. We outline the key components and methodology of this DRL-based approach.

State Space $\mathcal{S}$

The state s_t at time t includes information about the current beamforming state, the channel state information (CSI) $\mathbf{H}_t$, QoS requirements, and n_t.

$$s_t = (\mathbf{CSI}_t, \mathbf{H}_t, \mathbf{Q}_t, n_t)$$

where $\mathbf{Q}_t$ represents the QoS metrics at time t.

Action Space $\mathcal{A}$

The action a_t represents the decision to switch to a particular beamforming state s_j at time t.

$$a_t = s_j \quad \text{for} \quad j \in \{1, 2, \ldots, m\}$$

Reward Function $\mathcal{R}$

The reward r_t at time t is a function of the switching time and the QoS metrics. A higher reward is given for actions that minimize the switching time and maintain or improve QoS.

$$r_t = -\tau(s_i, s_j) + \sigma \cdot \text{QoS improvement}$$

where σ is a weighting factor balancing the trade-off between minimizing switching time and maintaining QoS.

Policy π_θ

The policy $\pi_\theta(s_t)$, parameterized by θ, maps the state space to the action space and is learned using the PPO algorithm.

Objective Function

We aim to learn a policy π_θ that maximizes the expected reward accumulated over time.

$$\max_\theta \mathbb{E}\left[\sum_{t=0}^{T} \gamma^t r_t\right]$$

where γ is the discount factor.

4 Steps to Integrate PPO Algorithm

Proximal Policy Optimization (PPO) is a widely used reinforcement learning algorithm valued for simplicity, stability, and efficiency. It enhances training stability by clipping the objective function, which limits large updates. PPO is also known for its sample efficiency, ease of parallelization, and straightforward implementation, establishing it as a trusted benchmark in reinforcement learning [10]. To integrate the PPO algorithm into the problem, we follow these steps:

4.1 Define the Policy Network

The policy network π_θ maps from the state space $\mathcal{S}$ to the action space $\mathcal{A}$. For PPO, this typically involves:

- `Input`: A set of state vectors with the shape corresponding to the concatenated state vector $s_t = (\mathbf{S}_t, \mathbf{H}_t, \mathbf{Q}_t, n_t)$.

- `Hidden Layers`:
  - Layer 1: Fully connected layer with 128 units and ReLU activation function.
  - Layer 2: Fully connected layer with 64 units and ReLU activation function.
  - Layer 3: Fully connected layer with 32 units and ReLU activation function.
  - Batch Normalization: Applied to stabilize training by normalizing the inputs to each layer.
- `Actor (Policy Network)`: The network output is a probability distribution over discrete beamforming actions, achieved through applying a softmax activation function to ensure the output values sum to one and represent probabilities.
- `Critic (Value Network)`: The network generates a single scalar output that represents the state value for the value function $V(s_t)$, using a linear activation function to ensure that the output remains untransformed.

4.2 Exploration and Exploitation

Generate beamforming using the current policy:

- `Interaction with Environment`: The base station (agent) applies the current policy π_θ to interact with the environment and gather a series of beamforming patterns. These patterns include states, actions, and rewards accumulated over time, aiming to balance the exploration of new beam configurations with the exploitation of established optimal actions to enhance overall performance.

4.3 Compute Rewards and Advantages

Calculate rewards and advantages from the collected beamforming:

- `Reward Function`: For each time step t, the reward function is computed as described earlier.
- `Advantage Function`: We use Generalized Advantage Estimation (GAE) to reduce variance in advantage values [9]. The `collect_beams` function selects actions using the policy network, interacts with the environment, and collects the resulting data.

4.4 Update Policy

Update the policy using PPO's clipped objective function $L^{\text{PPO}}(\theta)$:

- `Clipped Objective`:

$$L^{\text{PPO}}(\theta) = \mathbb{E}_t \left[\min \left(\psi, \text{clip}(\phi) \right) \right]$$

- $$\psi = \frac{\pi_\theta(a_t \mid s_t)}{\pi_{\theta_{\text{old}}}(a_t \mid s_t)} \hat{A}_t$$
- $$\phi = (\frac{\pi_\theta(a_t \mid s_t)}{\pi_{\theta_{\text{old}}}(a_t \mid s_t)}, 1 - \epsilon, 1 + \epsilon)\hat{A}_t$$

where $\hat{A}_t$ is the advantage at time t, $\pi_{\theta_{\text{old}}}$ is the previous policy, and ϵ is a clipping parameter.

– `Optimization`: The *Adam optimizer* is used to perform gradient ascent on the PPO objective, updating the policy network.

5. Repeat

`Iterate:`Continue the process by using the updated policy to collect new beamforming patterns, compute rewards and advantages, and iteratively update the policy.

6. Hyperparameter Tuning

Fine-tune the PPO hyperparameters, including the learning rate, batch size, and clipping parameter ϵ, to optimize performance. The hyperparameter values are detailed in the table below (Table 1):

Table 1. Hyperparameters for PPO

Hyperparameter	Value
Learning Rate	1×10^{-4}
Batch Size	64
Clipping Parameter (ϵ)	0.2

By implementing Proximal Policy Optimization (PPO), we aim to find an optimal policy π_θ that maximizes the cumulative reward while managing the trade-off between minimizing switching time and maintaining Quality of Service (QoS). PPO's clipping mechanism helps ensure stable and reliable policy updates, which is crucial for effective reinforcement learning.

5 Simulation and Evaluation

This section details the simulation environment and procedures employed to evaluate the proposed solution.

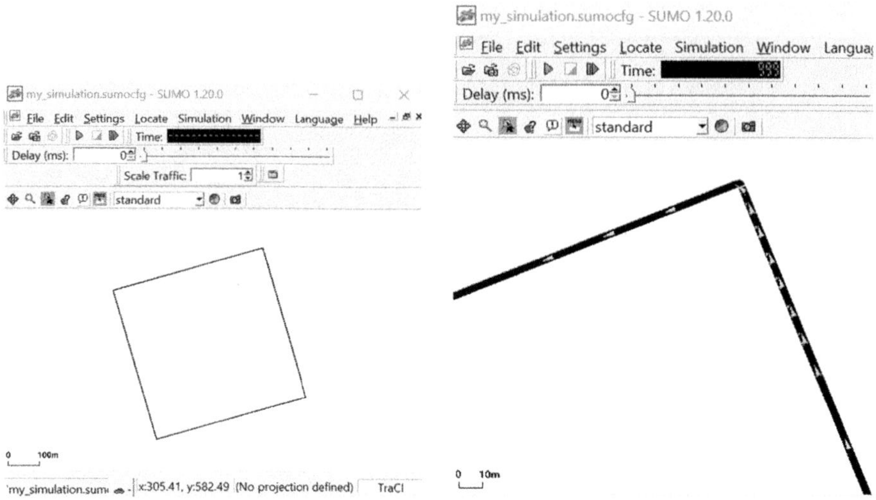

a b

Fig. 1. Overview of the Simulation in the SUMO Tool. (a) Zoomed-Out view (b) Zoomed-In view

5.1 Simulation Environment

The simulation environment models a base station with multiple antennas and various beamforming states for different channel conditions. It uses the SUMO (Simulation of Urban MObility) traffic simulator (Fig. 1) [1], integrated with a Deep Reinforcement Learning (DRL) agent. The DRL agent makes real-time decisions, while the simulation collects data on switching times and Quality of Service (QoS). Key 6G-V2X parameters include a frequency range of 2 GHz to 40 GHz and an antenna count of 64. The implementation was executed using Python version 3.9.19 (Fig. 1a and 1b).

The performance of the DRL model is evaluated against baseline methods, including a heuristic method that employs a fixed beamforming strategy and a traditional optimization method based on linear programming techniques. For 6G and V2X systems, maintaining robust communication requires meeting SINR thresholds of 10–15 dB for acceptable performance and data rate thresholds ranging from 5 Mbps for basic services to over 100 Mbps for advanced applications [3,4]. These thresholds are essential for ensuring high QoS in dynamic environments. In this evaluation, we empirically set the value of σ to 0.05, while α is set to 0.9 and β to 0.001.

5.2 Dataset

We utilized the NEU dataset, which is essential for developing and evaluating algorithms for hybrid beamforming, mmWave communication, and channel estimation in massive MIMO systems [11]. This dataset, originally containing raw data such as channel state information (CSI) and received signal strength indicators (RSSI), was adapted to meet the specific needs of 6G-V2X systems. We enhanced it by creating a tailored dataset that includes beamforming patterns, CSI, QoS metrics, and antenna configurations within the 6G frequency spectrum.

5.3 Baseline Comparison

The results in Fig. 2 show that the PPO model performed better than both the heuristic and traditional methods, achieving an average reward of 75.0, compared to 60.0 for the heuristic and 65.0 for the traditional method. This indicates that the PPO model learned and adapted well to the environment, optimizing rewards more effectively. In contrast, the heuristic method, with its fixed strategies, did not adapt as well to changing conditions. Although the traditional method was better than the heuristic, it still didn't match the PPO model's performance, likely because it depended on fixed constraints that didn't fully address the environment's complexities.

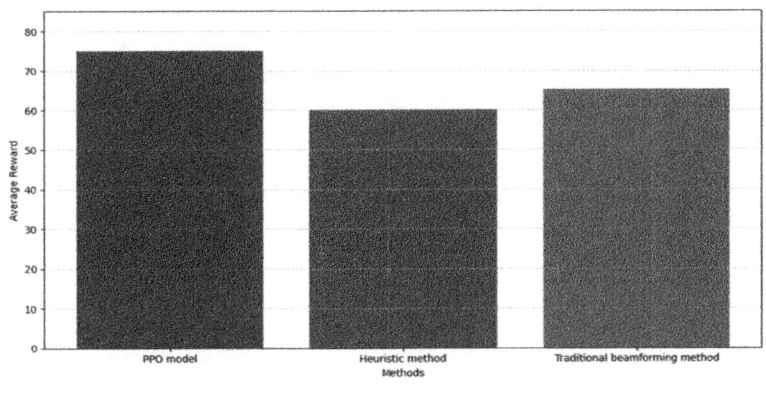

Fig. 2. Description of the problem under study

5.4 Effectiveness in Reducing Switching Time

From the results shown in Figs. 3, 4, and 5, the DRL model achieves an average switching time of 0.529 ms, significantly outperforming both the Traditional Beamforming method (0.546 ms) and the Heuristic-Based methods (0.560 ms). This highlights the DRL model's superior efficiency in adapting beamforming directions to vehicle movements, which translates to improved communication quality and reduced latency. These findings indicate that DRL is highly effective in optimizing beamforming in dynamic environments, thereby enhancing overall network performance for real-world applications.

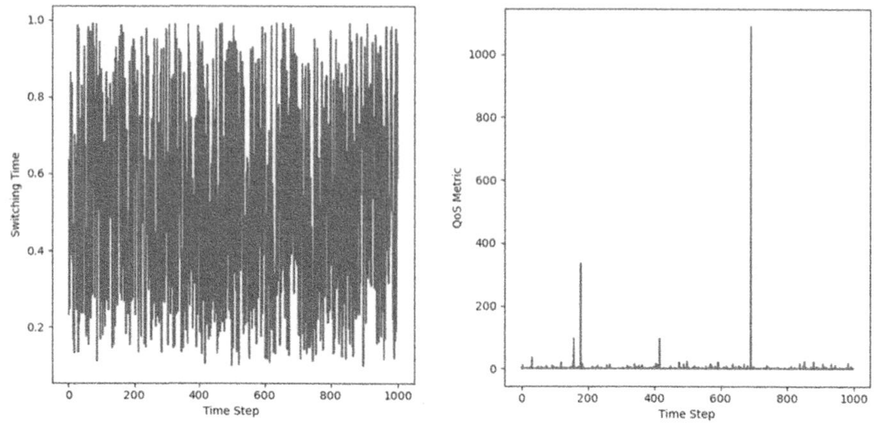

Fig. 3. Switching Time and QoS Results for PPO

5.5 Effectiveness for QoS

The analysis of the QoS results, as shown in figures (3, 4, and 5), reveals that the PPO model, with a QoS score of 3.419, performs better than the traditional and heuristic-based methods, which score 0.899. The higher QoS score for the PPO model indicates its effectiveness in meeting the required SINR and data rate thresholds essential for robust V2X communication. In contrast, the traditional and heuristic-based methods struggle to consistently maintain these thresholds, resulting in less reliable communication.

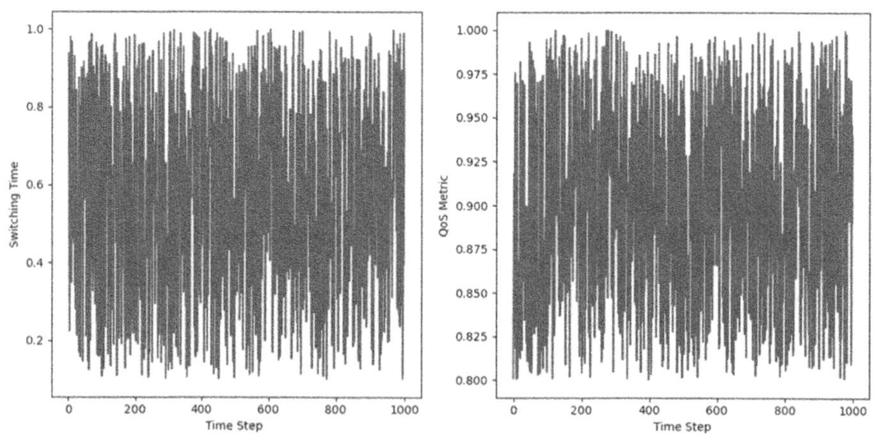

Fig. 4. Switching Time and QoS Results for heuristic-based method

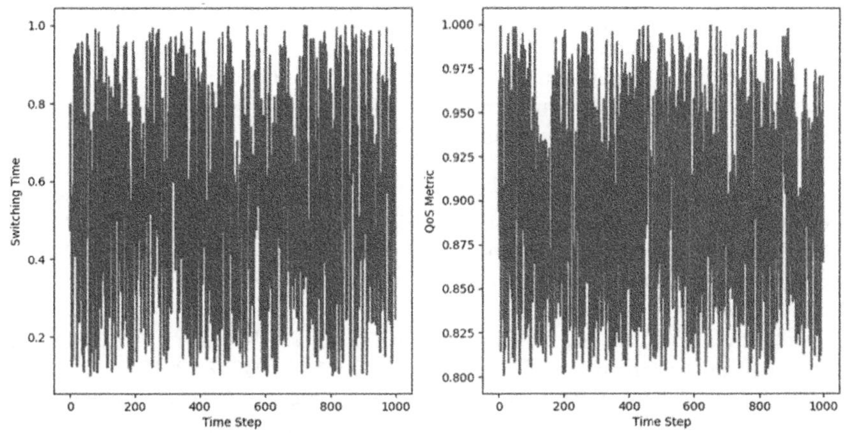

Fig. 5. Switching Time and QoS Results for traditional beamforming method

6 Conclusion

The simulation and evaluation demonstrate that the DRL model effectively minimizes beamforming switching time while maintaining QoS, highlighting its potential for real-world deployment in 6G-V2X communication systems. The DRL-based approach, utilizing Proximal Policy Optimization (PPO), successfully balances exploration and exploitation, enhancing beam switching time and service quality. Although the DRL model is highly effective, further improvements could be made by exploring hybrid approaches or advanced traditional methods. Adopting the PPO model is recommended, with additional research to enhance robustness and include standardized metrics for better comparison and user experience assessment.

References

1. TraCI: traffic control interface (2024). https://sumo.dlr.de/docs/TraCI.html. Accessed 26 Aug 2024
2. Ahmed, T.H., Tiang, J.J., Mahmud, A., Gwo Chin, C., Do, D.T.: Deep reinforcement learning-based adaptive beam tracking and resource allocation in 6g vehicular networks with switched beam antennas. Electronics **12**(10), 2294 (2023)
3. Boban, M., Manolakis, K., Ibrahim, M., Bazzi, S., Xu, W.: Design aspects for 5G V2X physical layer. In: 2016 IEEE Conference on Standards for Communications and Networking (CSCN), pp. 1–7. IEEE (2016)
4. Cavalcante, A.M., et al.: Performance evaluation of LTE and Wi-Fi coexistence in unlicensed bands. In: 2013 IEEE 77th Vehicular Technology Conference (VTC Spring), pp. 1–6. IEEE (2013)
5. Chen, K., Liu, D., He, X.: Fast beam switching based on machine learning for MmWave massive MIMO systems. In: Wang, X., Wong, K.-K., Chen, S., Liu, M. (eds.) AICON 2021. LNICSSITE, vol. 396, pp. 19–29. Springer, Cham (2021). https://doi.org/10.1007/978-3-030-90196-7_2

6. Luong, N.C., et al.: Applications of deep reinforcement learning in communications and networking: a survey. IEEE Commun. Surv. Tutorials **21**(4), 3133–3174 (2019)
7. Moltchanov, D., Beschastnyi, V.A., Ostrikova, D., Gaidamaka, Y., Koucheryavy, Y., Samouylov, K.E.: Optimal antenna locations for coverage extension in sub-terahertz vehicle-to-vehicle communications. IEEE Trans. Wireless Commun. **22**(9), 5990–6002 (2023)
8. Noor-A-Rahim, M.: 6G for vehicle-to-everything (V2X) communications: enabling technologies, challenges, and opportunities. Proc. IEEE **110**(6), 712–734 (2022)
9. Schulman, J., Moritz, P., Levine, S., Jordan, M., Abbeel, P.: High-dimensional continuous control using generalized advantage estimation. arXiv preprint: arXiv:1506.02438 (2015)
10. Schulman, J., Wolski, F., Dhariwal, P., Radford, A., Klimov, O.: Proximal policy optimization algorithms. arXiv preprint: arXiv:1707.06347 (2017)
11. Northeastern University. Neu datasets (2024). https://repository.library.northeastern.edu/files/neu:m613mx563. Accessed 19 Auf 2024

Reducing BLE Energy Loss in Busy 2.4GHz Band

Morgane Joly[1(✉)], Éric Renault[2], and Fabian Rivière[3]

[1] Université Caen Normandie, ENSICAEN, CNRS, GREYC UMR 6072, 14000 Caen, France
morgane.joly@unicaen.fr

[2] LIGM, Univ. Gustave Eiffel, CNRS, ESIEE Paris, 93160 Noisy-Le-Grand, France
eric.renault@esiee.fr

[3] NXP Semiconductor, 14000 Caen, France
fabian.riviere@nxp.com

Abstract. Even when coexistence techniques are in place to reduce the impact of interference, over-consumption is inevitable due to packet collision resulting in multiple retransmissions. To ensure good signal quality, more flexible countermeasures are needed to allow communication protocols.

This paper proposes a study of the performance of a Bagged Tree to predict the optimal Low Noise Amplifier (LNA) Gain according to the previously received packets with reduced input data.

Our study shows that stable interferer power AGC index prediction improves by 10% points the packet error rate and 96% of the future bad reception packet can be detected.

1 Introduction

Data collection has become the cornerstone of the flexibilization of all kinds of sectors (automotive field [32], the agriculture field [6] to the medical field [26], construction field [8] and personal equipment). To realise the communication between all devices, wireless protocols are the preferred solution, however spectral resources are limited. As a result, the last few years have seen an increase of the proliferation of Internet of Thing (IoT) devices, which drastically reduce the available spectrum. The 2.4GHz ISM band, one of the most used bands, is specifically impacted, challenging all the protocol on their quality of service (QoS). Narrow-band Low power/Low-rate protocols like BLE – designed to optimise the energy resources consumed by Internet-of-Things (IoT) devices – are especially *concerned/threatened* by the QoS degradation.

Whilst these protocols are useful to energy optimisation in short distance communication contexts, their strategy weakens the reception within high wireless network activity radio frequency bands. The studies [10,30] highlight the latency increase during devices discovery procedure and the correlation with the number of interfering devices of the same protocol or large sensor networks are degraded by Wi-Fi.

F. Hacène et al. (Eds.): MLN 2024, LNCS 15540, pp. 85–96, 2026.
https://doi.org/10.1007/978-3-032-00552-6_6

In this paper we proposed an improvement of the Automatic Gain Control computation to improve the roots of the reception chain quality of signal. Since original Wi-Fi signal power can be access directly, one of the easiest ways to reach an insight of these informations is to analyse quality of signal metrics generated during the packet reception that conceal the spectral environment impacts on the BLE packet.

The aim of this paper is twofold. The first one lies in the prediction of Automatic Gain Control for future BLE packets, the second consists in the prediction of future unfavourable reception slot.

The rest of the paper is organised as follows. Section 2 presents some related works. Section 3 introduces needed knowledge on BLE receivers, Sect. 4, describes the data collection and the feature extraction process and Sect. 5 where the results are discussed. We conclude in Sect. 6.

2 Related Work

The improvements allowed by the CR can be grouped in two main objectives :

- Improvement through analysis, where the targets use software-defined spectrum techniques. They are used to improve spectrum detection performance, cooperative detection and geo-localisation. The main challenge with these techniques is to make accurate decisions about spectrum availability, detection time, detection frequency and uncertainty in ambient noise.
- Improvement by radio reconfiguration, where the developed algorithm finality is to dynamically adjust the metrics parameters to improve the network performance. The main challenges are the complexity of the algorithm, the limited convergence time and the ratio between the energy consumption allocated to the algorithm and the energy gain allowed by the algorithm.

The intelligent channels assignment searches to improve the Adaptative Frequency Hopping with usually a channel whitelisting or blacklisting. The differentiation of the types of interference impact would allow to propose an efficient connection offset to allow the master-slave pair to synchronise their exchanges between WLAN bursts or, as realised by Hanna et al. [14], to facilitate the construction of the bit-by-bit signal to reduce latency by reducing the number of packet retransmissions and power consumption. Hithnawi's thesis [15] proposes an improvement to the error detection algorithm using the Hamming distance and RSSI with empirical thresholds. It does not use AI algorithms and the improvements in error detection are very promising. The Hamming distance quantifies the noise immunity, so it reflects the confidence in symbol decod. For each decoded symbol, it is quite eligible as a parameter in an ML algorithm.

Natively, devices cannot detect nearby protocol to apply the best countermeasure associated to them.

In the study [28], Sudharsan et al. proposes to add flexibility in the choice of the protocol used by the radio with a Support Vector Regression (SVR) that predicts the RSSI evolution to guide the choice of the best protocol in the future environmental conditions. Nevertheless, it implies integrating in the hardware all the possible protocols to allow a device to have the best choice panel.

Wang and Zhang in [31] use a decision tree to take in consideration the MAC layer in the spectrum sensing scheme to design a new strategy of emission and improve the quality of service of secondary users by wisely choose the instant of emission to avoid primary user interference.

The work [11] on LSTM to predict spectrum hole is the most comprehensive that we have found on the subject. The team developed an LSTM to be trained with time-series datasets sensed from the electromagnetic environment to exploit the cyclic aspects of the spectrum to identify and predict the evolution of vacant portions or frequency holes of the radio spectrum in Cognitive Radio (CR) systems. However, given the time and accuracy needed to predict 500 samples – between 31.25 ms to 7.33 s depending on the model and the hyper-parameters tested – in semi-hosting i.e. with "infinite" resources access (personal computer with a 3.0 GHz, i7-9700F CPU, 32-GB RAM and NVIDIA® GeForce GTX™1650 SUPER GPU with 1,280 CUDA® nuclei) is not suited BLE resources constrains.

In [3], Audéoud and al. develop a classification model based on the Link Quality Indicator (LQI) that classifies 802.15.4 radio links in three categories: reliable, uncertain and weak links. Their approach emphasises the importance of checking connectivity in both directions to ensure reliability, requiring only a basic handshake. The counter-measure system which results are to avoid weak links unless necessary for maintaining connectivity. The study also suggests that this classification system can be applied to technologies like BLE.

3 Radio Frequency Background

BLE operates on one of the most used frequency bands in IoT the ISM band – 2.4 GHz – it is affected by the presence of other wireless technologies such as Wireless local area networks as Wi-Fi.

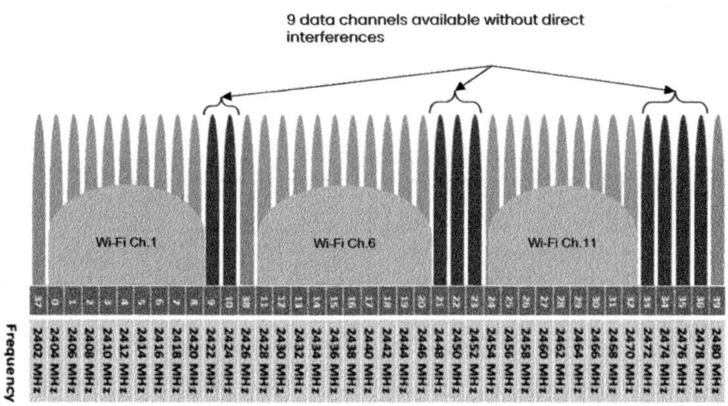

Fig. 1. Overlap of BLE by WiFi in ISM band.

In the case BLE radio system, only 3 advertising channels and 9 data channels amongst 39 are potentially free from direct disturbance coming from Wifi concurrent operations (see Fig. 1). However, due to intermodulation products that create perturbations in neighbouring channel, channels *a priori* free of any disturbance see their quality of signal decreased. Intermodulation products proportional to the interferes level during the amplification by the low noise amplifier (LNA) and which will add interference on wanted channel.

Several internal countermeasures mechanisms exist to attenuate the interferers effects. A tradeoff is made between energy saving and filtering complexity (Fig. 2).

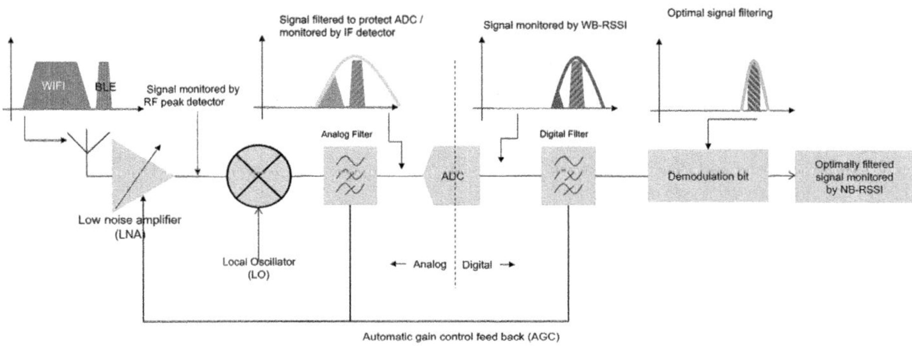

Fig. 2. Diagram of the radio and the different filters.

The AGC is used to avoid saturation during the payload reception by adjusting the receiver gain according to the received signal strength. The index will be set at the beginning of the packet reception, during the preamble, and frozen during the payload to avoid corruption of the bit decoding.

The AGC is therefore not set to sustain an increase of signal power during the payload reception that occurs when an interferer level creates receiver saturation (see Fig. 3).

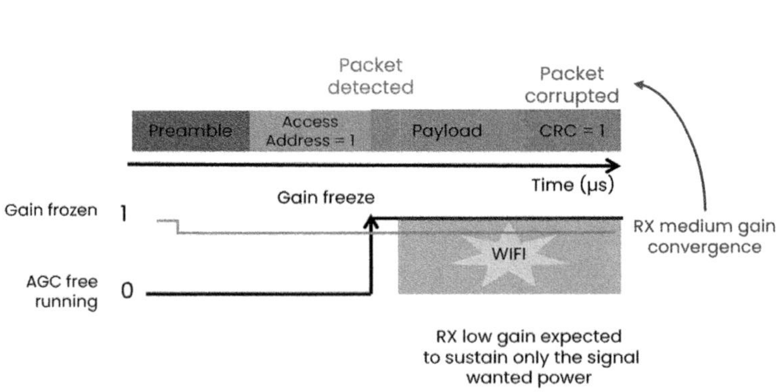

Fig. 3. Interferer arriving after the AGC freeze, perturbing the payload reception.

The aim of this study is to reduce the impact of such late arrival interferer by predicting the evolution of the interferer power on the packet reception without accessing directly to this information. Internal metrics that assess the quality of the signal are used to access indirectly to the evolution of the interferer:

- The Access Address Found event that allows the receptor to confirm the beginning of the reception.
- The Received Signal Strength Indicator that indicates the overall power in the monitored channel. It also adjusts the chain gain to guarantee the safe operation of the receiver.
- The SNR gives an indication of the signal purity.
- The LQI gives an indication on the solidity of the radio link.
- At the end of reception, a CRC word allows to detect errors during payload reception.

4 AGC Index Prediction for Future Packet Methodology

4.1 Delimitation and Scope

The experimental setup utilizes a BLE receiver model developed using Matlab tools, specifically Simulink and Matlab. The experimental plan focuses on the reception of BLE packets with signal power levels ranging from –100 dBm to 0 dBm, while being subject to interference from a Wi-Fi signal with power levels between –120 dBm and 0 dBm. This interference occurs at a frequency offset of 12 MHz from the BLE channel, which is chosen to represent the minimum separation between the desired signal and the interferer, thereby creating the most critical interference scenario. The power ratio between the desired and interfering signals is then varied during the experiment.

We have selected the bagged tree among the twenty-four algorithms proposed by the ML Graphical User Interface (GUI) of Matlab and according to our previous works on the analysis of ML algorithms [16].

4.2 Dataset Construction

Simulations are time-consuming and computationally expensive, so to streamline the data set generation, the signal experimentation plan is divided into smaller, manageable sections.

At the antenna, the signal under study is a combination of a BLE signal and a Wi-Fi interferer. The Wi-Fi signal appears intermittently, resulting in a pattern of BLE packets that are either interfered with or free from interference. Consequently, the BLE packet collection is categorized into two groups: packets with interference and packets without interference. To simplify the model, we assume the interferer is always present. For the second group, the interferer's power ranges from –120 dBm to –91 dBm, while for the first group, where the signal level is assumed to be too weak to detect with the interferer, the interference level ranges from –90 dBm to 0 dBm.

The interferer may appear either at the start or after the AGC index freeze (see Fig. 3). The resulting metrics collected during BLE packet reception are then used by machine learning (ML) algorithms to predict the optimal AGC index.

The repeatability of the experimental conditions is ensured by the control of the simulation seed for each simulated packet. Two datasets are produced, one with packets with the interferer arriving before the AGC index freeze and one with packets with the interferer arriving after the AGC index freeze.

Each packet reception scenario exists for the two versions of interferer arrival time.

Eventually, since the final application has a memory footprint constraint, we have chosen to limit data collection to three-time events spread across packet reception. Metrics that provide unique information during packet reception, such as Access Address Found and CRC, are stored at the beginning and end of packet reception respectively.

4.3 Packet Labelling

In Sect. 3, the role of the AGC mechanism in the receiver saturation avoidance and its limitation in case of late interferer arrival scenario has been exposed.

To label the data, we use the AGC index produced by the convergence of the AGC algorithm, retrieved during the simulation for each packet/interferer configuration according to the reception status of each packet. In most cases, the *native* AGC index value is considered, however some packets cannot be received due to the late interferer arrival.

We based the labelling on a logical array to determine the *Packet reception status* metrics. It combines both the **Access Address found** and **CRC** features (see Table 1). Three new determinations are created: *Good Reception (GR)*, *Bad Reception (BR)*, *No Detection (ND)*.

Table 1. Creation of packet reception status label

	Good reception (GR)	Bad reception (BR)	No detection (ND)
CRC	1	0	0
AA detected	1	1	0

The AGC index from packets received in same BLE power/Wi-Fi power conditions with different interferer arrival time can be compared according to the packet's reception condition. Five situations happen (the first two letters indicate the reception status for the version when the interferer arrives *Before* the AGC index freeze, while the last two letters indicate the reception status for the version when the interferer arrives *After* the AGC index freeze):

- GR_GR: Whatever the interferer's arrival time, the packet is always well received.
- GR_BR: If the interferer arrives late, the packet is received poorly because the AGC index is not adjusted sufficiently to avoid saturation.
- BR_BR: Whatever the timing of arrival of the interferer, the packet is never well received.
- ND_GR: The receiver expects a packet, and the interferer is too powerful for the receiver to detect the access address. However, the payload may be well received if the interferer arrives after the AGC index freeze.
- ND_BR: The interferer is too powerful for the receiver to detect the access address and the payload cannot be well received.

As a result, we label the data as described in Table 2.

Table 2. The determination of the packet class

Before_After	Label selected
GR_GR	AGC index from *Before* version
GR_BR	AGC index from *Before* version
ND_BR	X
ND_GR	X
BR_BR	X

GR: Good Reception, BR: Bad Reception, ND: No Detection

At the end of the labelling process, there are 12 numeric prediction classes in the range of 0 to 11, corresponding to the possible values of the AGC gain index, and one non-numeric value (X) to indicate that the future packet cannot be well received. There are therefore 13 prediction classes in total.

4.4 Model Evaluation

The true performance can only be revealed by the analysis of the packet status reception. For this second step of our analysis, a 10-packet sliding window Bagged-tree is evaluated online with a PER test closely related to the PER test used to examine the performance of the radio. The standard validation test consists in, for each Wi-Fi signal power, ranging from −100 dBm to 0 dBm, receiving 1800 packets at a fixed wanted signal strength. From this test, a bathtub curve is obtained allowing to characterise the performance of the receiver. The results are compared again to the 30.8% conformance PER established in the Bluetooth standard.

We have develop a new metric evaluating the proportion of well predicted bad packet named Future Bad Received rate (FBRR): when a packet is predicted to be badly received (i.e. the result of the Bagged Tree is X), the AGC algorithm converges without intervention: the packet is received with the AGC_{native}. The effectiveness of the prediction is then evaluated and quantified by the metric Future Bad Received rate (FBRR). It is the proportion of packets which are predicted as they cannot be well received during the experimentation.

5 PER Real Time Analysis

The full dataset is composed of 24 000 packets, the ratio between before and after packet is set within 0% to 100% range with a step of 30%. The train and test set have a wanted signal power within [–69, –19] dBm range and a interferer power within [–90, –5] dBm range. The model is trained with a packet history composed of the tenth last packet. The global accuracy of the obtain Bagged tree model is 69% ± 0.5%.

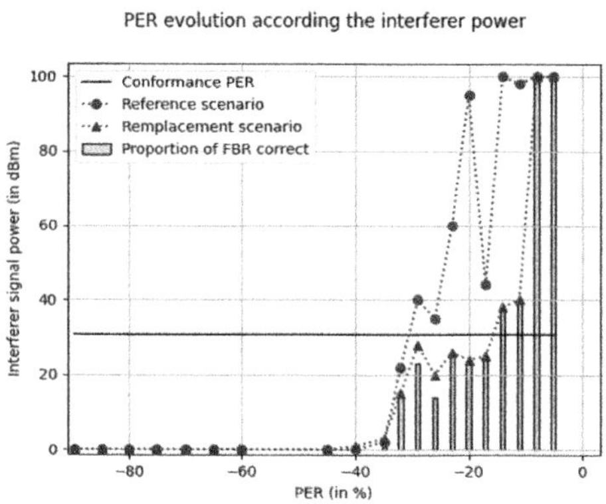

(a) Scenario 1.

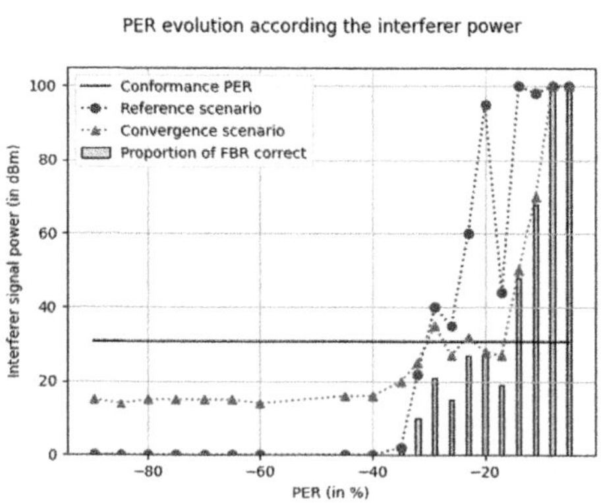

(b) Scenario 2.

Fig. 4. Packet Error Rate and Future Bad Reception Rate confirmed vs. the interference power with Signal power at -60dBm at 12 MHz Offset Reference scenario compared with ML-based scenario.

The model is evaluated on the simulated test bench to make real time prediction. We have considered two scenarios of AGC predicted use, the first one bypass totally the native AGC (see Fig 4a). The other set the maximum bound of the native AGC convergence, limiting the ability of the native algorithm to select high index if the model predict a future interference rise while the AGC algorithm in free running mode still adjust further the RX gain if needed (see Fig 4b). Each scenario is compared to the performance of the reference scenario in blue and the conformance PER sets á 30.8%.

In Fig. 4a the PER of the by-pass scenario is compared to the reference scenario. This scenario presents a PER with 10% points less than the other scenario, bringing the total PER below the conformance PER unlike the reference scenario. It can be highlighted that it temperate the peak of error between −29 dBm and −17 dBm, th model does not degrade the receiver performance low interferer context. Additionally, the majority of the bad predicted/ no detected packet has been predicted by the ML, 96% of the BR/ND packet are predicted as such.

In Fig. 4a the PER of the convergence scenario is compared to the reference scenario. This scenario presents a PER with equivalent performance compared to the reference scenario. Its results are interesting for medium to powerful interferer, it also temperate the peak of error between −29 dBm and −17 dBm. It performances are degraded for low interferer context. This scenario could be activated only when the interferer is detected as active.

Table 3. Comprehensive results of all scenario tested

Scenario	PER Mean	MER	MEC	FBR
Reference	31.8% ± 22.5%	/	−1.0182% ± 22.5%	/
By-pass	21% ± 17.4%	10.8% ± 11%	9.8% ± 17.4 %	96%
Range convergence reduction	30.6% ± 15%	1.2% ± 12.3%	0.2% ± 15%	67%

MER: Mean error with reference scenario, MEC: Mean error with conformance PER.

6 Conclusion

The BLE is a low power protocol optimized to save energy compared to other protocol as Bluetooth. It is subject to many sources of sensitivity degradation from Wi-Fi and other unregulated power technologies on near by channel. This leads to overconsumption, throughput decrease and connection failure.

This study of shown that on simulated results, the bagged tree prediction performance showed better result whit medium interfering Wi-Fi compared to the native algorithm, especially by by-passing totally the native AGC index that reduce by 10% points the PER and

The model is also able to detect future unfavourable slot and could allow to detect highly perturbed channel that could be use by the frequency hopping algorithm.

7 Further Research

The proposed method could save energy by avoiding unnecessary reemission, but the ML prediction is more consuming than the native algorithm, energy balance needs to be evaluated to determine the limit of interest of the method.

This study explores only the Bagged tree use with 10 packets sliding window, its performance needs to be compared with these of other models with various sliding window.

References

1. Ieee recommended practice for information technology– local and metropolitan area networks– specific requirements– part 15.2: Coexistence of wireless personal area networks with other wireless devices operating in unlicensed frequency bands. IEEE Std 802.15.2-2003 (2003). 1–150
2. Androec, D.: Using internet of things to tackle covid-19: a systematic review. In: 2021 IEEE 3rd Eurasia Conference on Biomedical Engineering, Healthcare and Sustainability (ECBIOS), pp. 230–233 (2021)
3. Audéoud, H.-J., Heusse, M.: Quick and efficient link quality estimation in wireless sensors networks. In: 2018 14th Annual Conference on Wireless On-demand Network Systems and Services (WONS), pp. 87–90 (2018)
4. Barac, F., Gidlund, M., Zhang, T.: Lped: channel diagnostics in wsn through channel coding and symbol error statistics. In: 2014 IEEE Ninth International Conference on Intelligent Sensors, Sensor Networks and Information Processing (ISSNIP), pp. 1–6 (2014)
5. Chung, J., Park, J., Kim, C.-K., Choi, J.: C-scan: wi-fi scan offloading via collocated low-power radios. IEEE Internet Things J. **5**(2), 1142–1155 (2018)
6. de la Concepcion, A.R., Stefanelli, R., Trinchero, D.: Adaptive wireless sensor networks for high-definition monitoring in sustainable agriculture. In: 2014 IEEE Topical Conference on Wireless Sensors and Sensor Networks (WiSNet), pp. 67–69 (2014)
7. Duong, M.: Whitehoodhacker, 2021
8. Eberle, D.: Baumschlager eberle
9. Elhachmi, G.: Cognitive radio spectrum allocation using genetic algorithm. Springer (2016)
10. Elshabrawy, T.: Throughput analysis of IEEE 802.15.4 enabled wireless sensor networks under wlan interference. In: 2014 IEEE Fourth International Conference on Consumer Electronics Berlin (ICCE-Berlin), pp. 467–469 (2014)
11. Enwere, P.I., Cervantes-Requena, E., Camuñas-Mesa, L.A., de la Rosa, J.M.: Using anns to predict the evolution of spectrum occupancy in cognitive-radio systems. Integration **93**, 102070 (2023)
12. Eu, Z.A., Lee, P., Tan, H.-P.: Classification of packet transmission outcomes in wireless sensor networks. In: 2011 IEEE International Conference on Communications (ICC), pp. 1–5 (2011)
13. Grimaldi, S., Mahmood, A., Gidlund, M.: Real-time interference identification via supervised learning: embedding coexistence awareness in iot devices. IEEE Access **7**, 835–850 (2019)

14. Hanna, S., Dick, C., Cabric, D.: Combining deep learning and linear processing for modulation classification and symbol decoding. In: GLOBECOM 2020 - 2020 IEEE Global Communications Conference, pp. 1–7 (2020)
15. Hithnawi, A.: Low-power Wireless Systems Coexistence, 2016
16. Joly, M., Renault, É., Rivière, F.: Comparison of AI-based algorithms for low energy communication. In: Renault, É., Mühlethaler, P. (eds.) Machine Learning for Networking. MLN 2022. LNCS, vol. 13767, pp. 1–11. Springer, Cham (2023). https://doi.org/10.1007/978-3-031-36183-8_1
17. Kikuzuki, T., Wada, A., Hamaminato, M., Ninomiya, T.: Automatic standard classification method for the 2.4 GHZ ism band. In: 2017 IEEE 85th Vehicular Technology Conference (VTC Spring), pp. 1–5 (2017)
18. Kolakowski, M.: Automatic radio map creation in a fingerprinting-based ble/uwb localisation system. IET Microw. Antennas Propag. **14**(14), 1758–1765 (2020)
19. Lee, W.-H., Ozger, M., Challita, U., Sung, K.W.: Noise learning-based denoising autoencoder. IEEE Commun. Lett. **25**(9), 2983–2987 (2021)
20. Madyatmadja, E.D., Nindito, H., Dian Sano, A.V., Purnomo, A., Haikal, D.R., Sianipar, C.P.: Application of internet of things in smart city: a systematic literature review. In: 2021 1st International Conference on Computer Science and Artificial Intelligence (ICCSAI), vol. 1, pp. 324–328 (2021)
21. Musa, N.S., Mirza, N.M., Ali, A.: Current trends in internet of things forensics. In: 2022 International Arab Conference on Information Technology (ACIT), pp. 1–5 (2022)
22. Nikoukar, A., Shah, Y., Memariani, A., Gunes, M., Dezfouli, B.P.: Predictive interference management for wireless channels in the internet of things. In: 2020 IEEE 31st Annual International Symposium on Personal, Indoor and Mobile Radio Communications, pp. 1–7 (2020)
23. O'Mahony, G.D., Harris, P.J., Murphy, C.C.: Investigating supervised machine learning techniques for channel identification in wireless sensor networks. In: t2020 31st Irish Signals and Systems Conference (ISSC), pp. 1–6 (2020)
24. O'Mahony, G.D., Harris, P.J., Murphy, C.C.: Detecting interference in wireless sensor network received samples: A machine learning approach. In: 2020 IEEE 6th World Forum on Internet of Things (WF-IoT), pp. 1–6 (2020)
25. Palit, R.: Anatomy of wifi access traffic of smartphones and implications for energy saving techniques. Int. J. Energy
26. Rivero-Angeles, M., et al.: Mobile clustering scheme for pedestrian contact tracing: the covid-19 case study. Entropy **23** (2021)
27. Singh, P., Raj, G., Kumar, A.: A review: smart farming using internet of things and machine learning. In: 2023 10th International Conference on Computing for Sustainable Global Development (INDIACom), pp. 1124–1129 (2023)
28. Sudharsan, B., Breslin, J.G., Ali, M.I.: Adaptive strategy to improve the quality of communication for iot edge devices. In: 2020 IEEE 6th World Forum on Internet of Things (WF-IoT), pp. 1–6 (2020)
29. Tanwongvarl, C., Chantaraskul, S.: Performance comparison of learning techniques for intelligent channel assignment in cognitive wireless sensor networks. In: 2015 Seventh International Conference on Ubiquitous and Future Networks, pp. 503–507 (2015)
30. Treurniet, J.J., Sarkar, C., Prasad, R.V., De Boer, W.: Energy consumption and latency in ble devices under mutual interference: an experimental study. In: 2015 3rd International Conference on Future Internet of Things and Cloud, pp. 333–340 (2015)

31. Wang, D., Yang, Z.: An advanced scheme with decision tree for the improvement of spectrum sensing efficiency in dynamic network. In: 2016 9th International Congress on Image and Signal Processing, BioMedical Engineering and Informatics (CISP-BMEI), pp. 1288–1292 (2016)
32. Yao, W., Pang, Z., Zhuang, K., Shao, W.: Design and application of wireless temperature monitoring system for diesel locomotive in reconditioning field based on ZigBee network. In: Jia, Y., Zhang, W., Fu, Y. (eds.) Proceedings of 2020 Chinese Intelligent Systems Conference. CISC 2020. LNEE, vol. 706, pp. 470–478. Springer, Singapore (2021). https://doi.org/10.1007/978-981-15-8458-9_50

Leveraging SHAP to Advance the Robustness of Large Language Models

Mohamed Yacine Djema(✉), Hacene Fouchal, and Olivier Flauzac

LAB-I, University of Reims Champagne-Ardenne, Reims, France
mohamed-yacine.djema@etudiant.univ-reims.fr

Abstract. In the domain of natural language processing, adversarial attacks often exploit the importance of specific words to deceive models. Identifying these critical words is essential for developing robust defense mechanisms. The SHAP (SHapley Additive exPlanations) method provides a comprehensive approach to quantify feature importance in machine learning models. This paper discusses the advantages of integrating the SHAP method with the SelfDenoise technique, emphasizing how SHAP aids in pinpointing salient words that are typically targeted by attack strategies. Through this integration, we aim to enhance the model's resilience against adversarial perturbations by reinforcing the denoising process on words deemed significant by SHAP.

1 Introduction

The rise of large language models (LLMs) has revolutionized natural language processing, enabling advancements in tasks ranging from machine translation to sentiment analysis. Despite their impressive capabilities, these models are vulnerable to adversarial attacks—subtle perturbations to input data that can mislead the model into making incorrect predictions. Such vulnerabilities pose significant risks, especially in applications requiring reliability and security.

Traditional defense mechanisms against adversarial attacks, like adversarial training, involve retraining models on adversarial examples. While effective, these methods are computationally intensive and often impractical for largescale models. Alternatively, randomized smoothing has emerged as a promising technique to provide certified robustness. By adding noise to inputs and aggregating the model's predictions, this method creates a smoothed classifier that is more resistant to perturbations without the need for retraining.

Building upon randomized smoothing, recent work has introduced self-denoised smoothing specifically for LLMs. This approach leverages the inherent capabilities of LLMs to denoise corrupted inputs before making predictions, effectively enhancing robustness without additional training or external models. However, existing methods typically involve randomly masking portions of the input, which may not efficiently target the most vulnerable aspects exploited by adversarial attacks.

F. Hacène et al. (Eds.): MLN 2024, LNCS 15540, pp. 97–101, 2026.
https://doi.org/10.1007/978-3-032-00552-6_7

2 Related Works

Randomized Smoothing and Certified Robustness. Randomized smoothing has been proposed as a technique to transform any base classifier into a smoothed version that is robust against adversarial attacks. This method's core idea is to use noise to create a stable classifier whose predictions remain constant within some perturbation set. Cohen et al. (2019) [1] offered a tight analysis demonstrating how randomized smoothing under Gaussian noise provides certifiable robustness. Their work showcases state-of-the-art results on datasets like ImageNet, revealing that randomized smoothing can apply to large neural networks, a significant milestone compared to more traditional methods that struggle to scale in such high-dimensional settings.

The primary contribution of randomized smoothing lies in its simplicity and efficacy; it provides robustness guarantees without requiring intricate modifications of the base classifier. Instead, it leverages Monte Carlo algorithms for prediction and certification, thus accommodating various architectures, including those of considerable size, like ResNet-50 on ImageNet.

Self-denoised for Large Language Models. In another vein, Jiabao et al. (2024) [3] propose a novel method termed "self-denoised" to enhance the robustness of LLMs. Recognizing the limitation that LLMs typically perform sub-optimally on noisy inputs, they introduce a two-step process where the LLM first denoises the input before making predictions. This self-denoising approach leverages the multitasking capabilities of LLMs and resembles a denoising pipeline known in the computer vision community, yet with significant caveats—namely, it circumvents the need for external denoising models and additional training costs.

Self-denoised smoothing shows marked improvements in both empirical and certified robustness against adversarial attacks that target LLMs. Unlike prior methods like adversarial training, which often require full access to model parameters and impose high computational costs, self-denoised smoothing is practical and efficient. The framework has demonstrated superior performance on tasks that necessitate robust human alignment, such as preventing jailbreak attacks on LLMs, with minimal overhead.

3 Methodology

Our approach aims to use an explainability method for machine learning models : SHapley Additive exPlanations (SHAP) [5] in conjonction with the SelfDenoise method. SHAP provides a unified approach to explain the output of any machine learning model by considering the contribution of each feature to the prediction. It is based on Shapley values from cooperative game theory, which allocate payouts fairly to players based on their contributions to the total payout.

Rather than masking randomly p percent of words by the tag ¡mask¿ in the input text and thus producing multiple masked version of the input, we compute the SHAP algorithm on the input text to determine the most important words in term of contribution to the prediction of the model. The most important words will be masked first : in our method masking p percent of words means masking the top p percent of words.

Then we use the model itself to denoise the masked input. For that we simply prompt the model with the masked input and demand that it fills the masks thus producing a denoised text that keeps the meaning and semantics of the original text. Note that this step does not require any specific training on the denoising task.

Finaly we prompt the model to perform a classification task on the denoised text, compare the result to the ground truth and evaluate it's accuracy.

Masking the most important words rather than doing it randomly presents an advantage, because most of the attacks used in the realm of large language models, like DeepWordBug [2] and TextBugger [4], aims to modify the most influential words within a text to manipulate machine learning models' outputs. By identifying these key words, we can understand which parts of the input are most influential and potentially vulnerable to manipulation.

4 Experimental Results

To evaluate the effectiveness of our proposed method we tested it using the ALPACA model [7] wich is fine-tuned from a 7B LLaMA model, on the Agnews [8] dataset. In a futur upate we aim to also apply our model on the SST-2 dataset [6] and comparing it to SelfDenoise [3] method. Note that the ALPACA model was loaded using FP-16 precision because of constraints regarding the VRAM usage. As the mask rate increases from 0.1 to 0.9, there is a noticeable decrease in accuracy. Initially, at lower mask rates, the accuracy remains relatively high, around 0.85. However, as the mask rate progresses beyond 0.5, accuracy begins to decline more steeply, reaching around 0.45 at a mask rate of 0.9. This trend suggests that higher mask rates negatively impact accuracy, likely due to the excessive masking of input data affecting model performance.

Since we mask the most important words first the accuracy should drop significantly even with a small mask-rate but we do not observe that as shown in Fig. 1. This can be explained because the context that remains with the unmasked words is enough for the model to denoise the masked input and retain enough semantic meaning for the input to have the same sense.

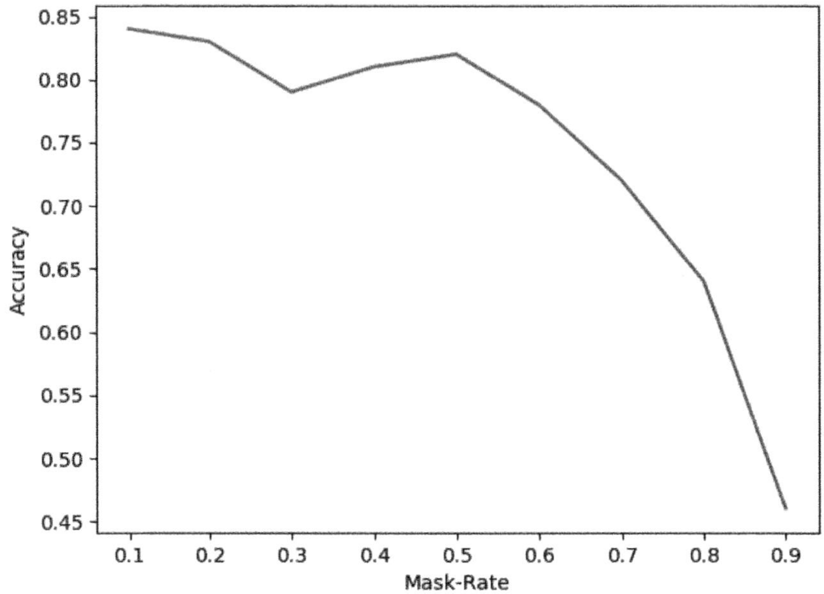

Fig. 1. Impact of Mask Rate on Accuracy

5 Conclusion and Perspective

In this paper, we introduced a novel approach to enhancing the robustness of Large Language Models (LLMs) by leveraging SHapley Additive exPlanations (SHAP) in conjunction with the self-denoised smoothing method. Our primary innovation lies in strategically masking the most influential words in an input text—as identified by SHAP—before prompting the model to denoise the input and perform classification. This technique contrasts with traditional randomized smoothing methods that randomly mask portions of the text without considering the contribution of individual words to the model's prediction.

Our experimental results using the ALPACA model on the AG News dataset demonstrate that our method maintains high accuracy even when significant portions of the most important words are masked. Specifically, the accuracy remains relatively stable at lower mask rates and only begins to decline noticeably as the mask rate exceeds 0.5. This resilience suggests that LLMs possess a robust capacity to recover semantic meaning from context, even when critical information is initially obscured.

By focusing on the most influential words, our method aligns with the tactics of adversarial attacks such as DeepWordBug and TextBugger, which target key words to manipulate model outputs. Understanding and countering these vulnerabilities is crucial for enhancing model robustness. Our approach not only serves as a defense mechanism against such attacks but also provides insights into the internal workings of LLMs by highlighting which words significantly impact predictions. Looking forward, we plan to extend our evaluation to additional datasets like SST-2 and conduct comparative analyses with existing methods, including SelfDenoise.

References

1. Cohen, J., Rosenfeld, E., Kolter, Z. : Certified adversarial robustness via randomized smoothing. In: International Conference on Machine Learning, pp. 1310–1320. PMLR (2019)
2. Gao, J., Lanchantin, J., Soffa, M.L., Qi, Y.: Black-box generation of adversarial text sequences to evade deep learning classifiers. In: 2018 IEEE Security and Privacy Workshops (SPW), 50–56. IEEE (2018)
3. Ji, J., et al.: Advancing the robustness of large language models through self-denoised smoothing. arXiv preprint arXiv:2404.12274 (2024)
4. Li, J., Ji, S., Du, T., Li, B., Wang, T. : Textbugger: Generating adversarial text against real-world applications. arXiv preprint arXiv:1812.05271 (2018)
5. Mosca, E., Szigeti, F., Tragianni, S., Gallagher, D., Groh. G. : Shap-based explanation methods: a review for nlp interpretability. 4593–4603 (2022)
6. Socher, R., et al. : Recursive deep models for semantic compositionality over a sentiment treebank. In: Proceedings of the 2013 conference on empirical methods in natural language processing, pp. 1631–1642 (2013)
7. Taori, R., et al. : Stanford alpaca: An instruction-following llama model (2023). https://github.com/tatsulab/stanfordalpaca
8. Zhang, X., Zhao, J., LeCun Y. : Character-level convolutional networks for text classification. Adv. Neural Info. Proc. Sys. **28** (2015)

Keyword-Driven Email Classification: Leveraging Machine Learning Techniques

Jiahui Xiang(✉), Osman Salem, and Ahmed Mehaoua

Centre Borelli, UMR 9010, Université Paris Cité, Paris, France
{jiahui.xiang,osman.salem,ahmed.mehoua}@u-paris.fr

Abstract. As companies grow larger through mergers and acquisitions, the number of identities within their organizational structures increases exponentially. These identities stem from diverse sources and serve various purposes, creating a significant challenge in effective classification. Generally, identities can be categorized into four main groups: Employees, who comprise the internal workforce; External Users, including clients and partners; Service Accounts, used for automated processes; and Generic Accounts, which allow shared access to resources.

To address this classification challenge, we propose a method that begins with identifying keywords associated with each identity category. This initial step is crucial for organizing identities. We then apply preprocessing techniques using lemmatization algorithms to standardize the terms, ensuring uniformity across different usages. Finally, we employ machine learning models to automate the classification process, improving the efficiency and effectiveness of identity management in large organizations. This comprehensive approach not only tackles the complexities of managing multiple identities but also enhances overall organizational security.

Keywords: Identity and Access Management · Identities · Classification models · email · lemmatization · Machine Learning

1 Introduction

Identity and Access Management (IAM) is becoming increasingly important in companies. An employee can have multiple accounts depending on their level of privilege, or even based on context. For example, an employee might be considered an internal user. However, in large companies, there are also more and more consultants, referred to as external users or contractors.

In addition to that, with the increasing number of connected devices, these devices must also have identities, often considered as service accounts. Depending on the company's structure, generic accounts are also created for specific purposes, such as reception or cash registers.

F. Hacène et al. (Eds.): MLN 2024, LNCS 15540, pp. 102–115, 2026.
https://doi.org/10.1007/978-3-032-00552-6_8

It is possible to classify identities based on the type of accounts:

- Employees
- Contractors
- Generic Accounts
- Service Accounts

The diversity and heterogeneity of these identities make their classification and identification difficult. This challenge is particularly prominent in older companies, due to the historical presence of legacy accounts. The technical evolution, addition of new functionalities, and, especially, mergers and acquisitions with other companies that have their own classification methods, all contribute to making the process of identity management increasingly complex, if not impossible.

This article proposes the use of classification methods involving machine learning and deep learning. By employing semi-supervised learning followed by self-supervised reinforcement learning, it would be possible to classify new accounts based on their name, surname, and email.

The remainder of this paper is structured as follows. Section 2 provides a review of recent work related to the detection of abnormal authentication delays in IAM. In Sect. 3, we introduce our proposed approach, which leverages lemmatization, rules to generate labeled data subsequently updates the classification model using the derived labeled data. In Sect. 4, we present the experimental results of our approach and compare the performance of supervised Machine Learning (ML) methods in terms of their detection accuracy. Finally, Sect. 5 concludes the paper and discusses potential future research directions.

2 Related Work

Bergmanis et al. in [1] present a novel sequence-to-sequence model for lemmatizing Universal Dependencies (UD) treebanks. Their approach leverages both sentence context and morphosyntactic features to disambiguate lemmas, making it particularly effective for morphologically rich languages where word forms can have multiple possible lemmas. The system addresses the challenge of handling ambiguous cases that require understanding the context, outperforming traditional lemmatization models by integrating contextual information directly into the model's predictions. By applying this method to various languages in the UD project, they demonstrate its robustness across diverse linguistic settings.

Zieliet al. in [10] propose a machine-learning classifier-based approach for language-independent lemmatization, focusing on the ability to scale across different languages. This model utilizes character-level embeddings along with a Random Forest classifier, showing that this method can adapt to both low-resource and morphologically rich languages. The system's success lies in its flexibility, being less dependent on language-specific rules or resources, which makes it highly applicable to languages with limited computational linguistic resources. They highlight how this approach achieves competitive performance and improved generalization across various linguistic datasets.

Bergmanis et al. in [2] examine a sequence-to-sequence learning model tailored for lemmatization, showing that such models can surpass traditional rule-based approaches, especially in languages with rich inflection. The model is designed to handle morphologically complex languages by learning word transformations directly from data. This paper focuses on how the sequence-to-sequence framework can learn lemmatization patterns without relying on hand-crafted rules. Their experiments demonstrate that this method is particularly effective for highly inflected words, providing a new way of thinking about lemmatization for complex languages.

Tiedemann et al. in [8] introduce a context-aware lemmatization model, which incorporates the surrounding words and morphosyntactic information into the lemmatization process. The authors tackle the challenge of lemmatizing words in morphologically rich languages, where traditional models struggle with homonyms and highly inflected word forms. By integrating context, the model achieves better disambiguation of lemmas, particularly in situations where word forms may have multiple valid interpretations. This paper demonstrates that context-aware models can outperform previous lemmatizers in several morphologically rich languages.

Kuhn et al. in [4] present a novel approach to lemmatization using Neural Machine Translation (NMT) models, where lemmatization is treated as a translation task. By translating word forms into lemmas using character-level embeddings and attention mechanisms, the authors show that NMT-based models can outperform traditional models, especially for languages with complex morphology. The paper emphasizes the advantages of using NMT architectures, including their ability to generalize to unseen word forms and handle the morphological richness of languages like Finnish and Russian. The results show that NMT models are a promising direction for future lemmatization systems.

Li et al. in [5] explore how BERT-based models can be adapted for language-specific lemmatization tasks. The authors fine-tune pre-trained BERT models to use their contextual embeddings for disambiguating lemmas. This paper demonstrates that BERT's ability to capture surrounding context and linguistic nuances can lead to significant improvements in lemmatization accuracy, particularly in ambiguous cases where traditional rule-based approaches fail. This research provides insights into how transformer-based models, like BERT, can be extended to handle morphological tasks in natural language processing (NLP).

Smith et al. in [7] propose a multilingual machine learning model for lemmatization, focusing on cross-lingual transfer learning. Their model is trained on several languages and learns to generalize lemmatization rules, enabling it to perform well on languages it has not been explicitly trained on. The authors demonstrate that by leveraging multilingual data, their system can significantly improve lemmatization for under-resourced LaTeX Warning: Empty 'thebibliography' environment on input line 6. languages. The paper highlights the potential of multilingual models to share linguistic knowledge across languages, making it a valuable tool for expanding NLP resources in low-resource settings.

Oliver et al. in [6] present a deep learning model for lemmatization that operates at the character level, capturing the morphological structure of words more effectively than word-level models. The authors show that character-level models can better handle highly inflected languages, where individual words may take on many different forms. By focusing on subword units, the model generalizes well to unseen word forms, providing superior performance in languages with complex morphological rules. This research highlights the benefits of using deep learning approaches for fine-grained morphological tasks like lemmatization.

Cotterell et al. in [3] introduce a joint learning model that tackles both morphological reinflection and lemmatization in a unified framework. By treating these two tasks as related, the authors show that their model learns to leverage information from each task to improve performance. This approach allows for better generalization and more efficient learning, particularly in morphologically rich languages where word forms undergo complex transformations. The joint model outperforms separate task-specific models, demonstrating the advantages of multitask learning for related linguistic tasks.

Zieliet al. in [9] explore decision tree ensemble methods for lemmatization tasks, comparing them to deep learning-based models. The authors use Random Forest classifiers to handle the lemmatization process, showing that decision tree ensembles are particularly effective for low-resource languages where training deep learning models may not be feasible. Their results indicate that decision tree ensembles can achieve competitive performance, often surpassing neural networks in specific scenarios, particularly when training data is scarce or computational resources are limited. This paper provides a strong argument for considering ensemble methods in NLP tasks, especially in low-resource contexts.

3 Proposed Approach

In the realm of natural language processing, particularly in email classification tasks, the initial challenge often arises from the fact that the data are not labeled. To effectively categorize the emails into distinct classes, it is essential to establish a dataset comprising relevant keywords for each class. This step ensures that the classification model has a basis for understanding the context of the emails. For each email in the dataset, we will focus on retaining only the top-level part of the email address—omitting the domain part—to streamline the data and enhance relevance. Once the data has been pre-processed, the next step involves splitting the text into individual components and performing lemmatization. This process simplifies words to their base or root form, which aids in reducing dimensionality and improving the model's efficiency. Subsequently, a keyword filter will be applied to the processed emails, labeling them according to their corresponding classes based on the presence of the predefined keywords. Following this labeling process, we will convert the emails into a numerical format by encoding them into integers. This transformation is vital for feeding the data into machine learning algorithms, which require numerical input for training. Ultimately, with the data pre-processed, labeled, and encoded, we can proceed to train the classification model and utilize it for predictions on new, unseen email data.

Lemmatization is a fundamental process in natural language processing (NLP) that involves reducing words to their base or root form, known as a lemma. Unlike stemming, which simply truncates words to their base form, lemmatization considers the context and converts the word to its meaningful base form. For example, the words "running" and "ran" are reduced to the lemma "run." This process is crucial for various NLP tasks, including information retrieval, text mining, and machine translation. Mathematically, lemmatization can be expressed as a function $f : W \to L$, where W is the set of all words in a given corpus, and L is the set of all possible lemmas. The lemmatization function can be formally defined as:

$$f(w) = \begin{cases} \text{lemma}_w & \text{if } w \in W \\ \text{undefined} & \text{if } w \notin W \end{cases}$$

This function f maps each word w in the set W to its corresponding lemma lemma_w in the set L.

Label encoding is a technique used in machine learning to convert categorical variables into numerical format. This transformation is essential for algorithms that require numerical input, as many machine learning models cannot directly interpret categorical data. In label encoding, each unique category is assigned an integer value. For instance, the categories "red," "green," and "blue" may be encoded as 0, 1, and 2, respectively. This process allows algorithms to treat categorical features numerically while retaining their distinct categories. Mathematically, label encoding can be represented as a function $g : C \to N$, where C is the set of unique categories and N is the set of integers. The encoding function can be defined as follows:

$$g(c) = i \quad \text{for } c \in C, \text{ where } i \text{ is the index of } c \text{ in the sorted list of } C.$$

This function g assigns an integer index i to each category c in the set C, effectively converting categorical data into a numerical format suitable for machine learning algorithms.

To classify emails effectively, we employ a keyword-based approach derived from manual analysis for each class. This method allows us to categorize emails into distinct classes, specifically Generic and External. By referencing the keywords outlined in Table 1 and Table 2, we ensure that emails are accurately classified according to their content and context. This structured classification process not only enhances our email management efficiency but also aids in filtering and prioritizing communications based on their relevance to our operations.

Identifying service accounts can be more complex compared to personal accounts due to their diverse naming conventions and patterns. However, there are several keywords and characteristics that can help distinguish service accounts from regular user accounts. First, certain keywords are strong indicators of service accounts. These keywords include:

Table 1. Generic

account	accueil	acunetix	adm	admin	alsid
amministrazione	application	astreinte	audithuman	beauty	billeterie
boardroom	booking	boutique	bureau	catalog	chauffeur
client	cockpit	comite	commande	commerce	communication
comunicazione	concierge	confezionamento	confirmaciones	consultation	contact
control	contrôle	corporate	customer	cyber	data
desk	digital	electronique	elettronica	emballage	enquiries
enquiry	evenement	expedition	expenses	experience	experiencia
exploitation	export	facturation	facture	fatturazione	fax
filiale	finance	fiscal	formation	google	holding
hostess	hq	hôtesse	infirmerie	info	information
infos	internet	intranet	invoice	ipad	iphone
livraison	logisti	magasin	magazzino	mailbox	maintenance
management	manager	marchandise	marketing	medico	monitor
notification	office	okta	operat	payable	payment
pickpup	planning.	prd	privacy	procurement	produccion
production	qualit	recepcion	reception	reception	recruitment
relation	reply	request	reserv	resource	ressource
restaurant	retail	room	rsvp	salesforce	screen
securit	service	servicio	sharepoint	shipment	shipping
signaletique	sms	staff	stock	store	superviseur
suport	support	system	tax	team	technolo
temp	test	textile	ticket	tivoli	tmp
training	travel	tst	usercube	vendor	warehouse
workplace	zscaler				

Table 2. External

ext.	ext-	.ext	-ext	-ext.
ext-	_ext	ext_	.ext.	-external
.external	external.	externe.	partner.	partners.
-externe	guest.	.guest	guest.	.guest.
	guest-	guest.		

- **service**: This is a straightforward keyword that explicitly indicates a service account.
- **svc**: A common abbreviation for service, this keyword often appears in service account names.
- **s-**: Prefixes like "s-" are frequently used in naming conventions for service accounts, making it an important indicator.

- **s**: The single letter "s" can be a part of various service-related identifiers.
- **Numerical Prefixes**: In some cases, service account names may start with a number, such as "1_service" or "2_svc". This can help differentiate them from user accounts, which are less likely to have numerical prefixes.

In addition to these keywords, it's also important to consider other coding patterns. For example, names that begin with a number, such as "1234-start", might suggest a service account, particularly in environments where numerical prefixes are commonly used to denote automation scripts or backend processes. By incorporating these keywords and patterns into our classification algorithms, we can improve the accuracy of identifying service accounts. This refined classification can help in better categorizing emails and enhancing the overall efficiency of the email management system.

For employee, we employ a straightforward method for determining employee identifiers when keywords are not available. If the identifier does not fall under the categories of generic, service, or external, we classify it as an employee. The following formats are used for constructing the employee's identifier: firstname.lastname, lastname.firstname, firstnameFirstLetter.lastname, and lastnameFirstLetter.firstname. This structured approach ensures consistency in our naming conventions and helps maintain clarity in our employee records.

4 Experimental Results

Table 3. Class Identification Index

Number	Category
0	employee
1	external
2	service
3	generic

We are leveraging a comprehensive dataset provided by a company, which encompasses over 300,000 accounts aggregated from 52 different systems into a robust Identity and Access Management (IAM) solution. Each account within this dataset is uniquely identified by its associated email address. The classification of these accounts is meticulously mapped as outlined in the accompanying table, facilitating a clear understanding of the various categories present within the dataset Table 3. This structured mapping enhances our ability to manage and analyze user accounts effectively, ensuring streamlined access control and identity verification processes.

All benchmarks are performed on the following machine:

- Machine: T480 (2020)
- CPU: i7-8550U (4 cores - 8 threads/min 1.8 GHz - max 4.0 GHz)
- Memory: 24 GB RAM

In our text processing workflow, we utilize the Natural Language Toolkit (NLTK) for lemmatization, which allows us to convert words to their base or dictionary form, enhancing the accuracy of our data analysis. After lemmatization, we apply filters based on keywords specified in Tables 1 and 2 to classify the accounts accordingly. Additionally, we conduct a search for service accounts by identifying keywords such as svc, s-, and service. This systematic approach ensures that we accurately categorize and manage accounts based on their specific characteristics and functions, contributing to more effective identity and access management. Then we apply multiple machine learning model.

Models are evaluated based on their performance in terms of accuracy, F1-score, recall, precision, Receiver Operating Characteristic (ROC) curve, and Area Under the Curve (AUC), as well as their time to train and predict. These metrics are defined as follows:

- Accuracy: The proportion of correct predictions to the total predictions made.
- Precision: The proportion of correct positive predictions.
- Recall: The proportion of actual positives that are correctly predicted.
- F1-score: A harmonic mean of precision and recall.
- AUC: A measure of the overall discriminative power of a model.
- Training Time: The time required to train a model.
- Prediction Time: The time required to predict the class of a new data instance.

We use models:

- **LR**: Logistic Regression
- **NB**: Naive Bayes
- **ET**: Extra Trees
- **RF**: Random Forest
- **MLP**: Multi-Layer Perceptron
- **DT**: Decision Tree
- **XGB**: XGBoost
- **LGB**: LightGBM
- **GB**: Gradient Boosting
- **AB**: AdaBoost
- **SV**: Support Vector Machine (See Figs. 7, 9, 10 and 11).

Highest Accuracy: The models **SV** - Fig. 2, **ET** - Fig. 1, **RF** - Fig. 3, **XGB** - Fig. 4, and **DT** - Fig. 8 have the highest accuracy scores, around 0.934–0.935, indicating strong performance in correctly predicting labels.

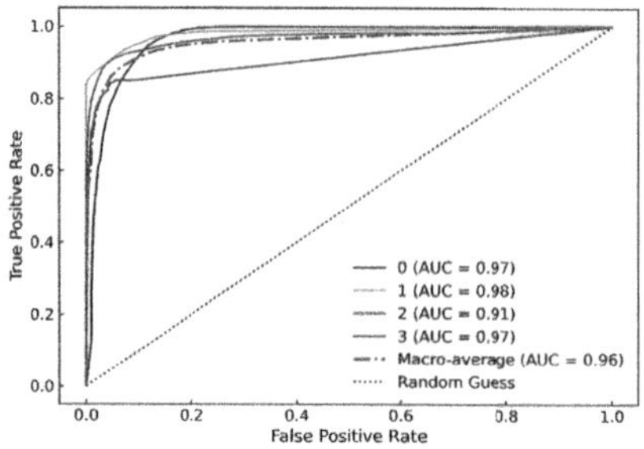

Fig. 1. extra trees

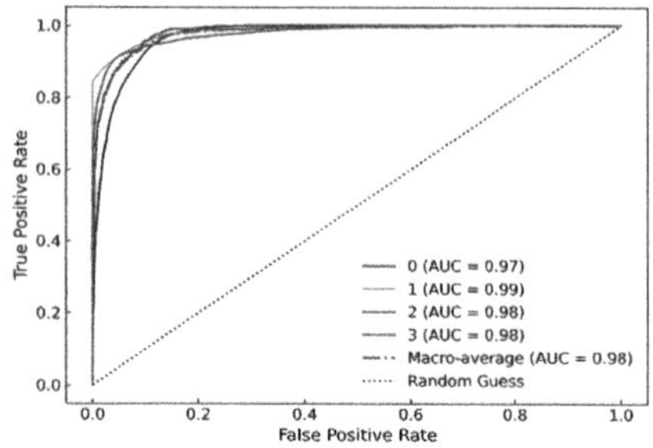

Fig. 2. svm

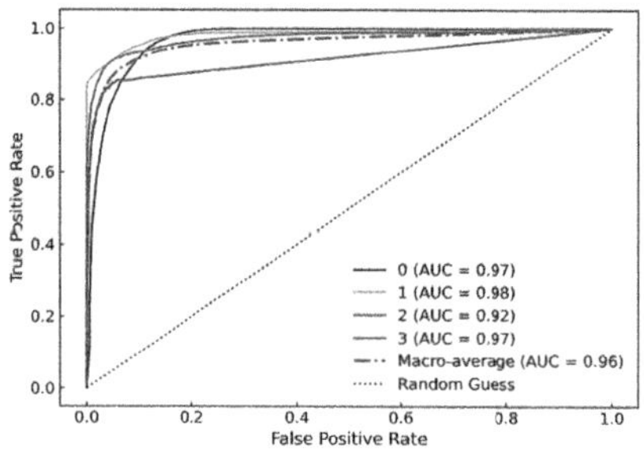

Fig. 3. Random Forest

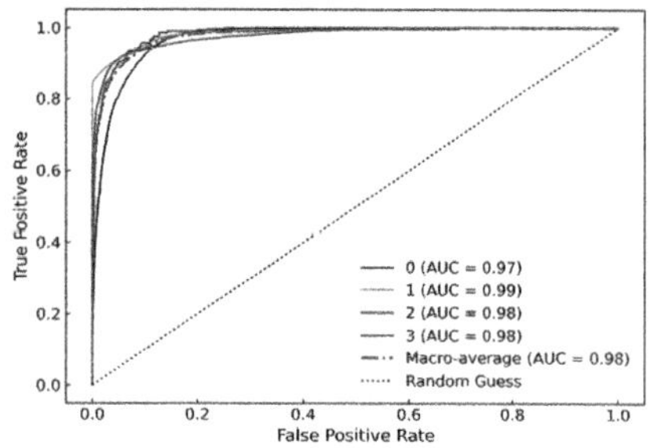

Fig. 4. eXtreme Gradient Boosting

Naive Bayes (NB): NB - Fig. 5 shows a lower accuracy (0.729), although it has a high precision (0.920), suggesting it makes fewer false positives but has a lower overall prediction rate.

Logistic Regression (LR): LR - Fig. 6 has the lowest scores across accuracy, recall, and F1-score, indicating it might not be well-suited for this dataset.

SV and XGB: Both SV - Fig. 2 and XGB - Fig. 4 perform particularly well in terms of precision and recall, maintaining a high F1-score, suggesting they are well-balanced in managing both false positives and false negatives.

Figure 12 shows Accuracy, F1-Score, Precision and Recall for each. Most models perform well, achieving high values across all metrics, with scores generally above 0.8 for each metric. XGB, RF, ET, and GB tend to have consistently high performance across all metrics, which is common for ensemble-based models. Naive Bayes (NB) generally shows lower performance, particularly in Recall and F1-score, compared to other models, suggesting it may struggle with this dataset. Support Vector Machine (SV) performs.

Figure 13 shows testing times for each model. Ultra-Fast Prediction Models: Most models, including LR, NB, DT, and RF, demonstrate extremely fast

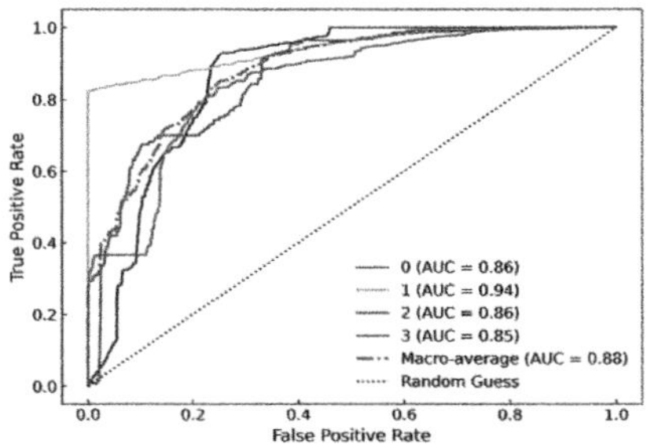

Fig. 5. Gaussian Naive Bayes

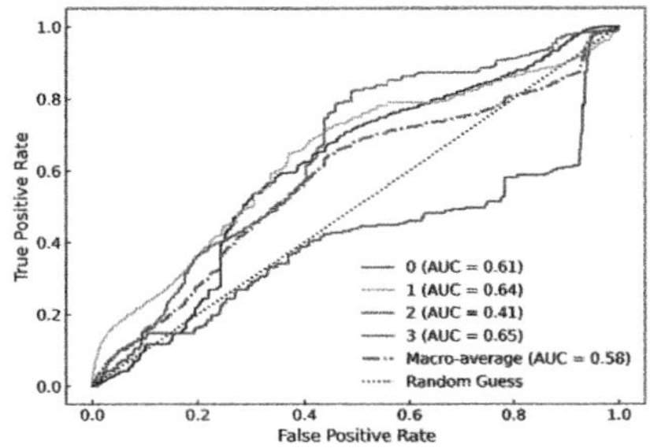

Fig. 6. Logistic Regression

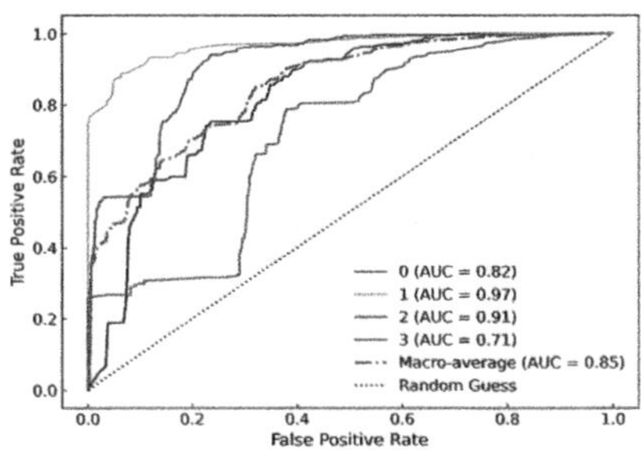

Fig. 7. Adaptive Boosting

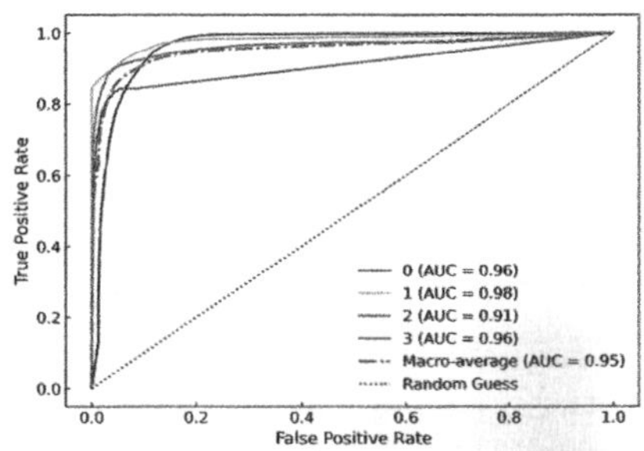

Fig. 8. Decision Tree

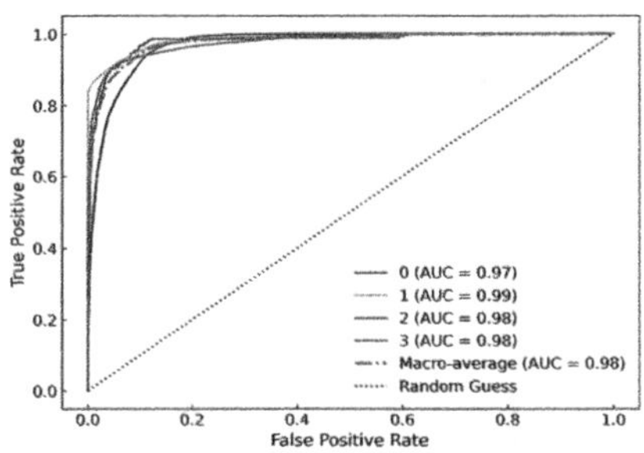

Fig. 9. Light Gradient Boost

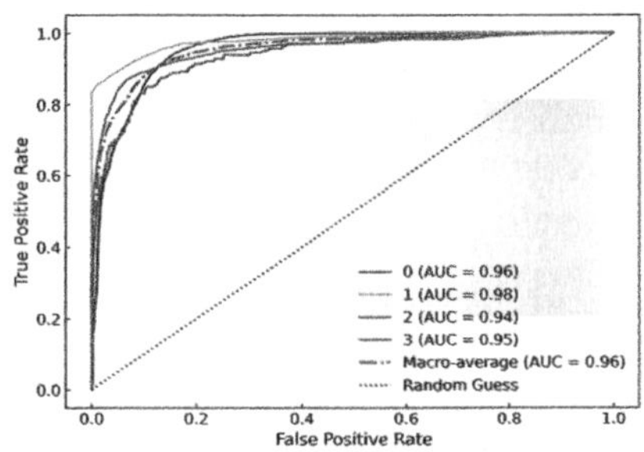

Fig. 10. Multi Layer Perception

prediction times, typically in the range of microseconds to nanoseconds. This makes them highly suitable for real-time applications where rapid predictions are essential, such as in online decision systems or streaming data scenarios. SVM (Support Vector Machine): The SV model has the longest prediction time, approximately 52 microseconds, which is slightly higher compared to other models. Although relatively slower, it remains efficient for many practical applications and offers a trade-off with its high accuracy and AUC performance. Ensemble Models (ET, RF, XGB, GB, and LGB): Despite their complexity, these ensemble methods show reasonable prediction times, mostly under 10 microseconds.

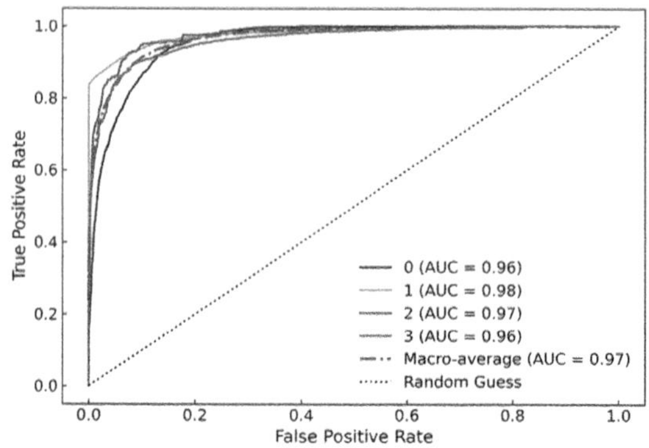

Fig. 11. Gradient Boosting

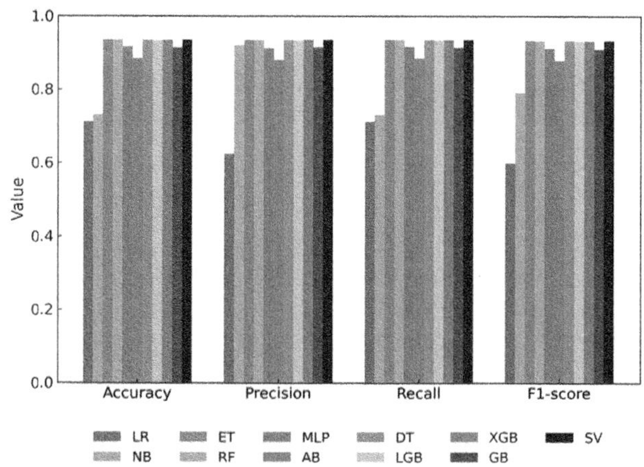

Fig. 12. Accuracy, F1-Score, Precision and Recall for each model

This efficiency can be attributed to optimized algorithms, especially in XGB and LGB, making them viable for high-frequency prediction tasks. AdaBoost (AB): With a prediction time of around 8 microseconds, AB is also quite efficient, suggesting it can be deployed in scenarios where quick responses are critical. MLP (Multi-Layer Perceptron): While it has a relatively longer training time, MLP has a fairly efficient prediction time at 6 microseconds, indicating that once trained, it can still make rapid predictions.

Figure 14 shows training time for each model. NB and DT models train in less 1 s. Then LR, ET, RF, AB, LGB and XGB perform around 5 s. GB and SV train took respectively 101 s and 54 s. The worst is MLP with 901 s (Table 4).

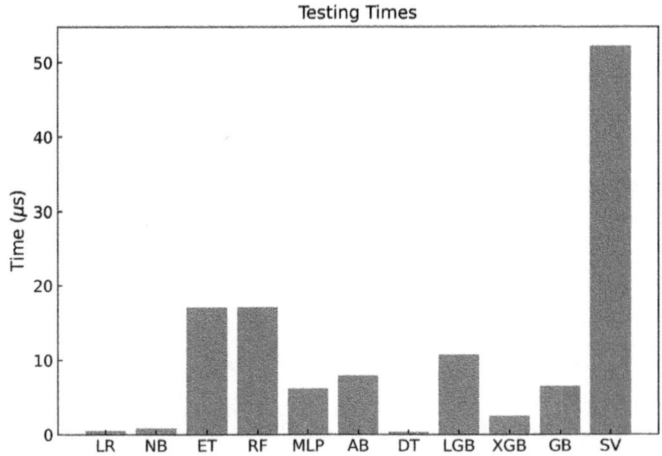

Fig. 13. testing times.

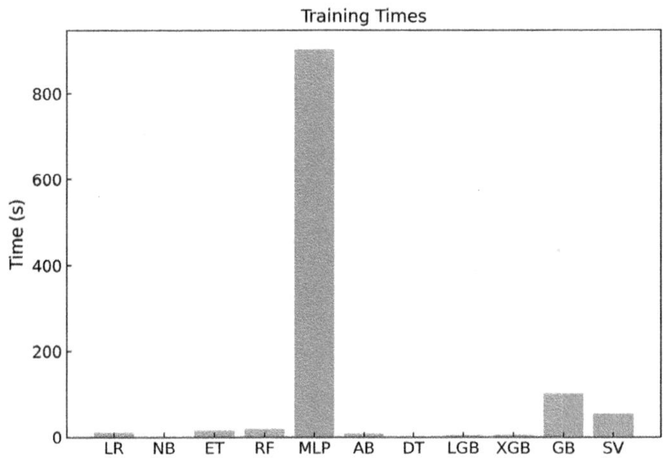

Fig. 14. training times

For High Performance: If computational resources are not a constraint, XGB and SV are excellent choices given their balanced metrics and high AUC. For Fast Deployment: NB and DT provide quick training and prediction capabilities with reasonable accuracy, suitable for time-sensitive applications. Balanced Choice: LGB and RF offer a good balance between speed and accuracy, with high AUC and relatively lower training time compared to MLP and GB.

Table 4. Summary

models	accuracy	recall	precision	f1	train time (s)	prediction time (s)	auc
nb	0.7293	0.7293	0.9199	0.7902	0.15135	8.0169e-07	0.8789
et	0.9346	0.9346	0.9343	0.9324	15.2523	1.7025e-05	0.9567
rf	0.9346	0.9346	0.9343	0.9324	18.5122	1.7068e-05	0.9598
mlp	0.9155	0.9155	0.9115	0.9114	901.425	6.1734e-06	0.9565
ab	0.8842	0.8842	0.8795	0.8782	7.88205	7.9032e-06	0.8534
dt	0.9340	0.9340	0.9335	0.9318	0.78486	2.9664e-07	0.9535
lgb	0.9330	0.9330	0.9330	0.9308	3.55049	1.0702e-05	0.9782
xgb	0.9341	0.9341	0.9344	0.9317	4.59340	2.4381e-06	0.9791
gb	0.9136	0.9136	0.9140	0.9093	101.170	6.4952e-06	0.9684
sv	0.9349	0.9349	0.9348	0.9327	54.4157	5.2156e-05	0.9790

5 Conclusion

In this paper, we present a solution that begins with keyword identification, followed by labeling through lemmatization techniques, and concludes with machine learning applications. Classifying accounts based on email addresses is crucial for effective identity management in large organizations, particularly those with subsidiaries or undergoing mergers. This classification facilitates improved segregation of privileges. A key element of this process is the identification of relevant keywords. Looking ahead, it would be valuable to incorporate an automatic detection system for new keywords using a semi-supervised learning approach.

References

1. Bergmanis, T., Goldwater, S.: Universal lemmatizer: a sequence-to-sequence model for lemmatizing universal dependencies treebanks. arXiv preprint arXiv:1902.00972 (2020)
2. Bergmanis, T., Goldwater, S.: Sequence-to-sequence learning for lemmatization. Nat. Lang. Eng. **27**(2), 187–214 (2021)
3. Cotterell, R., et al.: Morphological reinflection and lemmatization using joint learning. In: Proceedings of the 58th Annual Meeting of the Association for Computational Linguistics, pp. 3245–3256 (2020)
4. Kuhn, J., et al.: A neural machine translation approach to lemmatization. arXiv preprint arXiv:2004.03341 (2020)
5. Li, W., et al.: Bert-based models for language-specific lemmatization. arXiv preprint arXiv:2106.05502 (2021)
6. Oliver, S., et al.: Character-level deep learning for lemmatization. J. Artif. Intell. Res. **70**, 1–22 (2021)
7. Smith, J., et al.: Multilingual lemmatization with machine learning models. Comput. Linguist. **48**(3), 101–121 (2022)

8. Tiedemann, J., Scherrer, Y.: Context-aware lemmatization for morphologically rich languages. In: Proceedings of the Sixth Conference on Machine Translation, pp. 508–513 (2021)
9. Zieliński, T., et al.: Decision tree ensembles for lemmatization. Nat. Lang. Eng. (2021)
10. Zieliński, T., et al.: Highly language-independent word lemmatization using a machine-learning classifier. Revista Mexicana de Ingeniería Biomédica **42**(1), 26–41 (2021)

Predicting Intents: ARMA-Based Modeling

Nagham Hachem(✉) and Éric Renault

LIGM, Univ. Gustave Eiffel, CNRS, ESIEE Paris, 93162 Marne-la-Vallee, France
{nagham.hachem,eric.renault}@esiee.fr

Abstract. With the introduction of 5G technology, operators can now design systems that are scalable, effective and flexible thanks to previously unheard-of possibilities for virtualized and intent-based network management. Intent-based networking makes network administration easier while improving dependability and flexibility by concentrating on "what" results are sought rather than "how" they are attained. This study investigates intent prediction in 5G networks using Auto-Regressive Moving Average (ARMA) modeling. ARMA provides a linear, resource-efficient method for predicting important intent variables like bandwidth, CPU and memory needs by utilizing previous data. In situations with little datasets, ARMA's ease of use and effectiveness are demonstrated through comparison with machine learning methods. By adding external factors and non-stationary data trends, extensions such as ARMAX and ARIMA improve predictions even further. These findings illustrate ARMA's potential for proactive intent management, assuring optimal resource allocation in dynamic 5G contexts, and laying the framework for sophisticated machine learning techniques integration.

1 Introduction

5G technology is a revolutionary advancement in mobile networks, offering unmatched speed, low latency, and vast connectivity potential. It enables operators to build multiple virtual networks within a single physical infrastructure, promoting innovation and flexibility. Intent-based network management, combined with 5G technology, simplifies administration by focusing on the "what" rather than the "how." This approach ensures effective, error-free operations and uniform policy enforcement, making networks smarter, more robust and adaptive. This technology will open new avenues for development and economic growth.

2 Programming Languages

Programming languages [2,19] are essential for communication between computers and humans. They consist of instructions coded in a specific style, evolving from low-level machine language to high-level languages. These languages

F. Hacène et al. (Eds.): MLN 2024, LNCS 15540, pp. 116–132, 2026.
https://doi.org/10.1007/978-3-032-00552-6_9

have evolved over time, becoming more powerful and user-friendly. They can be divided into five generations based on performance, robustness and reliability (Fig. 1).

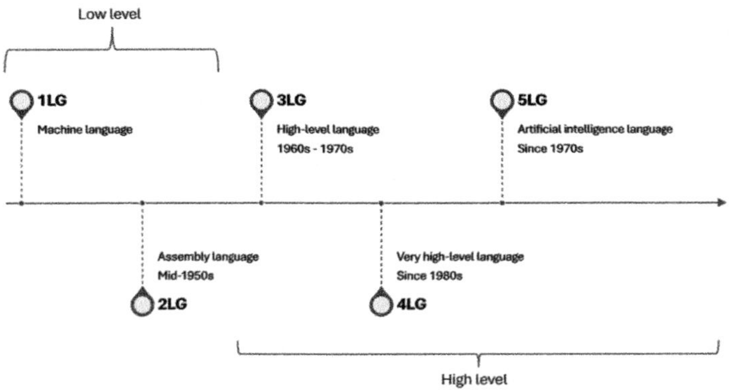

Fig. 1. The different generations of programming languages

2.1 Low Level

First-Generation Languages (1LG): Machine languages, starting in the early days of computing, are low-level binary languages that directly represent electrical states and instructions. They are fast and efficient but challenging to learn due to their complexity and lack of human readability.

Second-Generation Languages (2LG): Assembly languages emerged in the mid-1950s, providing a more human-readable format using mnemonic codes and symbols. They require an assembler to translate into machine code and are architecture-dependent, offering easier programming compared to machine languages.

2.2 High Level

Third-Generation Languages (3GL): Developed during the 1960s-1970s, these languages (like C, Pascal and Fortran) use English-like syntax and require a compiler to translate programs into machine code. They are more comprehensible, portable across different machines with specific compilers, and reduce the amount of code needed compared to earlier generations.

Fourth-Generation Languages (4GL): Introduced from the 1980s onward, 4GLs (such as SQL, Matlab, and R) are non-procedural and user-friendly, resembling natural languages. They focus on higher-level operations, allowing users to specify "what" needs to be done rather than "how," thus increasing productivity in tasks like database management and data analysis.

Fifth-Generation Languages (5GL): Originating in the 1970s, 5GLs use advanced concepts like artificial intelligence, constraint programming and parallel computing to solve complex problems. They aim for full artificial intelligence integration, improving automation and intelligence in applications, albeit at the cost of more intricate and expensive infrastructure requirements.

The development of programming languages, from binary simplicity to complex 5GLs, demonstrates significant advancements in computer science. These languages simplify the programming process, allowing users to define objectives in natural ways. Despite drawbacks like higher memory needs and less hardware control, the trend towards more widespread and efficient programming is strong.

3 What is Intent?

An Intent, as defined by organizations such as GSMA, ETSI, TM Forum, IETF, and 3GPP, as shown on Fig. 2, represents high-level directives focussing on network behavior and business objectives and guiding network configurations toward specific goals.

Depending upon the normalization institution, Intents have slightly different meanings:

- **GSMA** [5]: Intents are high-level, outcome-oriented assertions that guide network configurations and actions using AI and machine learning to simplify network management and automation.
- **ETSI** [10]: Intent is a specification of expectations, including requirements, goals and limitations, serving as a guideline for network management and orchestration, moving the focus from detailed configurations to overall network purposes.
- **IETF** [4]: Intent is a set of operational objectives expressed declaratively, enabling automatic network configurations in line with operator intents, improving efficiency, scalability and adaptability.
- **TM Forum** [7]: Intent focuses on system aims and business objectives, formally stating all expectations without specifying the implementation details.
- **3GPP** [1]: Intent is a high-level declaration of system goals, outlining expectations, requirements, goals, and constraints, which can be broken down into policies and activities for network management.

The characteristics of intents differ across standards bodies such as GSMA, ETSI, IETF, TM Forum, and 3GPP. Key features include a user-friendly interface, AI/ML automation and a customer-centered approach within 5G network slices. Some organizations prioritize anomaly detection and learning integration, while others focus on simplifying management and anticipating outcomes. Understanding these variances is crucial for navigating the evolving field of intent-based networking.

4 Intent Languages

Intent Languages are high-level programming languages that let administrators define intended results rather than specific configurations, which simplifies network management. By automatically converting human-readable goals into low-level network regulations and configurations, intent languages offer abstraction, automation, flexibility, and scalability by concentrating on the "what" rather than the "how". They provide dynamic adaptations to shifting network conditions and increase efficiency while lowering mistakes. By using this method, administrators can manage extensive network infrastructures with consistent and logical policy enforcement, guarantee high-priority traffic, and preserve application accessibility (Fig. 3).

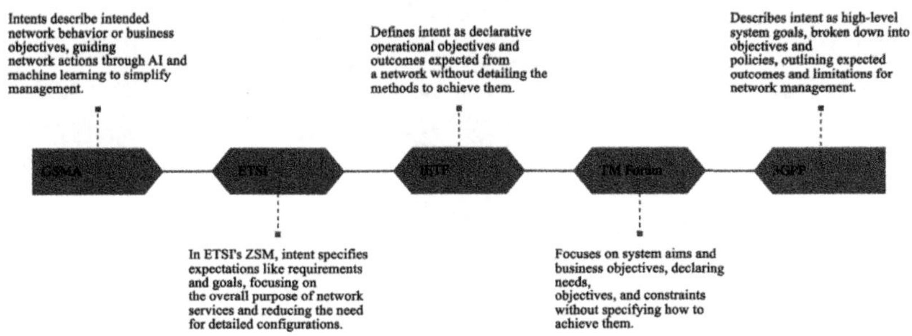

Fig. 2. Intent Definition Across Organizations

4.1 Domain Specific Language

Domain-specific languages (DSLs), as defined by [11], are programming languages designed for specific application domains or problem areas, offering higher abstraction levels and restricted syntax for easier representation and completion of activities.

- **Frenetic** [16] simplifies network programming by leveraging OpenFlow [16] and NOX [17]. It aims to abstract network complexities and reduce errors in managing and controlling network devices. Frenetic's architecture includes a runtime system for managing low-level switch rules and a high-level network query language, resembling SQL in structure. It offers abstractions to express network policies intuitively, hiding intricate network details from programmers.
  - **NetKAT** [14] was built upon Frenetic, grounded in Kleene algebra [23]. It provides a formal language for programming and reasoning about networks, characterized by denotational semantics mapping packet histories and equational theory ensuring program correctness. NetKAT's rule-based framework supports compositional network behavior definitions, enabling rigorous analysis and verification of network policies and behaviors.

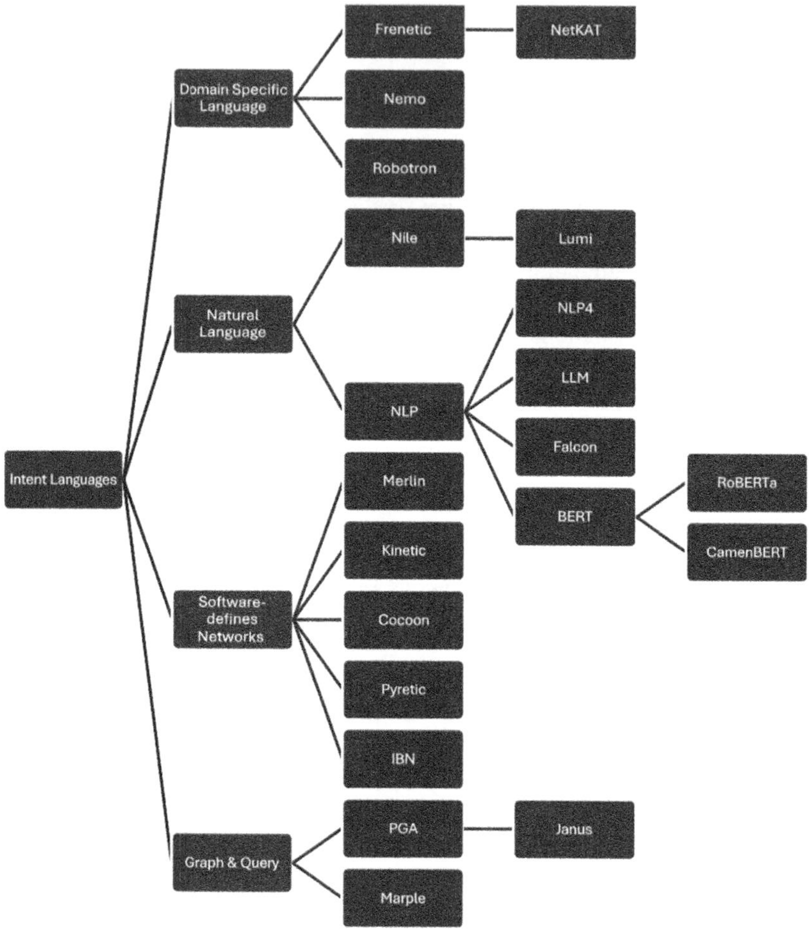

Fig. 3. Classification of Intent languages

- **Nemo** [41,42] is a domain-specific language (DSL) designed for intuitive network resource description. It abstracts network models and operations, focusing on user intent expression rather than low-level configurations. Nemo facilitates high-level intent modeling through structured object-operation-result (OOR) models, aligning with Intent-Based Networking (IBN) paradigms. It aims to simplify network management by allowing users to declare intent, leaving system complexities to be handled automatically.
- **Robotron** [39], developed by Facebook, automates network management tasks based on high-level operator intents. It uses FBNet to model network elements and configurations, supporting a top-down approach from design to deployment and monitoring. Robotron aims to minimize human involvement in network management by converting high-level intents into actionable configurations across large-scale networks. It emphasizes proactive management and efficient resource utilization through centralized network state management.

4.2 Natural Language

Natural languages [33], such as English or Standard Mandarin, are spoken by humans and are influenced by fundamental laws of all languages, according to the universal grammar hypothesis. Natural language processing studies these interactions.

- **Nile** [18] is a specialized language designed to express network management intentions clearly and abstractly. It bridges human-readable natural language intents with precise network configuration commands, ensuring completeness and correctness through enforced grammar. Nile's flexibility supports various network operations, making it accessible and manageable for network administrators and stakeholders alike.
  - **Lumi** integrates Nile's abstract tools to convert plain language queries into actionable network intents. It emphasizes interoperability and efficiency in translating human-centric intents into network configurations. Lumi's modular pipeline enhances accuracy from natural language input to network deployment, yet faces challenges in complex and ambiguous scenarios.
- **Natural Language Processing** (NLP) [6,8,15] enables computers to understand and manipulate human language, bridging computational linguistics with AI techniques like machine learning and deep learning. NLP systems handle ambiguity, syntactic analysis, and semantic understanding, crucial for applications like speech recognition, translation and conversational agents.
  - **NLP4** [15] translates high-level network intents from human language into P4 data-plane programs using NLP techniques and a MultiLayer Perceptron (MLP) model. It facilitates the execution of network actions from human-readable intentions, bridging the gap between intent specifications and network configurations.
  - **Large Language Models** (LLM) [20] exemplified by the GPT series, utilize transformer architecture to understand and generate human-like text. Trained on vast datasets, they excel in various NLP tasks but raise ethical concerns regarding biases and misuse, necessitating responsible deployment and ongoing research.
  - **Falcon** [37] aids in Named Entity Disambiguation (NED) by linking unstructured content to structured data in knowledge repositories like Wikidata. Falcon 2.0 enhances entity linking accuracy and scalability, supporting advanced NLP tasks with its robust framework and user-friendly interface.
  - **Bidirectional Encoder Representations from Transformers** (BERT) [22] revolutionized NLP by capturing bidirectional context and understanding complex language semantics. Pre-trained on large text corpora, BERT's transfer learning capabilities making it versatile for tasks like question answering, text classification and more.
    * **RoBERTa** [25] improves upon BERT with optimized pre-training and fine-tuning procedures for various NLP tasks. By enhancing data

processing, model design, and training strategies, RoBERTa achieves state-of-the-art performance in natural language understanding and generation.
* **CamemBERT** [26] is a French language model based on BERT, tailored for NLP tasks in French. It leverages transformer architecture for contextual comprehension of French text, achieving high performance through pre-training on extensive French corpora and fine-tuning for specific applications.

4.3 Software-Defined Networking

Software-defined networking (SDN) [9] revolutionized network design, operation and management by making networks dynamic, agile and adaptive. SDN integrates seamlessly into the cloud era, enabling dynamic setups that maximize performance and monitoring. It centralizes network knowledge through data and control planes, transforming traditional network architecture. SDN controllers streamline management procedures.

- **Merlin** [38] focuses on managing resources in Software-Defined Networks (SDNs) through a declarative programming language. It allows network managers to express high-level rules for forwarding pathways, bandwidth limits, and logical predicates. These rules are compiled into executable code, leveraging advanced compilation techniques and dynamic policy adaptation mechanisms. Merlin simplifies network administration by providing expressive abstractions and scalable policy management capabilities.
- **Pyretic** [31] is a domain-specific language for SDN programming that facilitates the definition of complex network rules and their composition. It abstracts away low-level details, enabling programmers to manage network behavior effectively. Pyretic integrates with SDN controllers and OpenFlow, translating high-level network policies into specific rules for network devices. Its declarative approach and modular design make it suitable for developing and deploying various SDN applications.
- **Kinetic** [21] is a flexible SDN controller system and domain-specific language designed for dynamic network operations. It uses an event-driven programming style and finite state machines (FSMs) to organize network policies. Kinetic includes automated verification capabilities using Computation Tree Logic (CTL) and NuSMV model checking to ensure the correctness of control algorithms. Built upon Pyretic, Kinetic emphasizes scalability, automated verification and ease of network control in both educational and practical settings.
- **Cocoon** [36] is an advanced framework for SDN development that integrates network design and verification processes. It employs a step-by-step refinement approach to modularize network designs and improve verification efficiency. Cocoon distinguishes between static network design verification and

dynamic configuration checks at runtime, enhancing overall network reliability. It supports automated verification and defect localization, making it suitable for complex SDN systems and real-world applications.

- **Intent-Based Networking** (IBN) [13] utilizes artificial intelligence (AI) and machine learning (ML) to automate network management based on high-level intentions expressed by administrators. It extends SDN by focusing on automating network behaviors in alignment with organizational goals. IBN features include configuration automation, troubleshooting, optimization and automatic provisioning, aiming to improve network agility, responsiveness and reliability. IBN represents a paradigm shift in network administration by reducing manual intervention and enhancing decision-making through analytics and proactive monitoring.

4.4 Graph and Query

A graph [3], consisting of vertices and edges, is a mathematical structure used to depict interactions between pairs of things in complex systems like social networks, communication networks and transportation networks. It is used in network policy management to visually describe rules and interactions between network entities. Queries [35], on the other hand, are requests for specific data or actions from a system or database, enabling network administrators to make informed decisions and ensure network operations to comply with standards.

- **Policy Graph Abstraction** (PGA) [34] revolutionized network policy administration by introducing a graph-based model for expressing high-level policies across diverse network environments. It addresses the complexities of modern networks, including mobility, security and performance requirements, by automating conflict detection and policy resolution. PGA's dynamic and Quality of Service (QoS) rules accommodate temporal aspects and performance constraints, making it a robust solution for managing policies in automated infrastructures like SDN and NFV environments.
  - **Janus** [12] extends the principles of PGA to offer a unified system for dynamic, group-centric policies in network management. It leverages heuristic methods for mobility and temporal dynamics, an Integer Linear Program (ILP) for group-based atomicity and bandwidth negotiation techniques for effective policy deployment. Janus integrates SDN programming languages like Pyretic to express middlebox functionality and evaluate service chains, emphasizing Intent-Based Policy Support and dynamic adaptation to diverse network conditions.
- **Marple** [32] is a performance query language designed for network performance monitoring, particularly targeting programmable switch environments. It supports functional programming features like groupby, zip, filter, and map to facilitate expressive queries for monitoring network performance metrics. Marple introduces a programmable key-value store primitive on switch hardware, enhancing scalability and efficiency in processing a large number of keys

required for comprehensive network monitoring. Its specialized compiler for P4-programmable software switches and high-speed simulator contribute to its practicality and broad applicability in network performance monitoring.

5 Reactivity vs. Proactivity

Reactive and proactive strategies are the two main strategies frequently used in management systems. Each provides a unique method for controlling next usage.

The main goal of a reactive approach is to react to the demand by using real usage data from the prior period. This method responds to demand as it emerges rather than attempting to forecast future requirements. When historical data alone offers enough insight into future trends, it is frequently employed. The reactive strategy may not always optimize usage since it lacks the foresight to adapt to variations in demand, even though it might be useful in stable circumstances.

In contrast, a proactive strategy uses historical data and real-time inputs to estimate future consumption. This approach makes it possible to better prepare and allocate resources by forecasting future consumption patterns, which results in more effective use. Because proactive techniques enable more flexible and anticipatory management, they are especially beneficial in dynamic contexts where demand fluctuates dramatically. Significant resource savings may arise from this foresight, particularly in situations when efficiency is crucial.

Collectively, these tactics reflect several management philosophies. The particular situation, the data at hand, and the intended results in terms of cost and savings, all play a significant role in the decision between reactive and proactive strategies.

6 Prediction

In many different fields, predictions are essential because they allow to predict future events based on the information that is currently accessible. Predictions can be made using a variety of methods, both linear and non-linear. Non-linear approaches work well with dynamic and non-linear systems and frequently make use of sophisticated models or machine learning algorithms.

In the following, we concentrate on linear forecasts, which are still essential because of their effectiveness and simplicity. In particular, we investigate their applications in time series analysis methods like ARMA (AutoRegressive Moving Average).

6.1 Linear Prediction

Linear predictions involve estimating future values using a linear relationship with past data points. They offer sophisticated and effective solutions by presuming that future trends can be represented as weighted combinations of earlier observations. Two excellent examples that provide strong instruments for evaluating and predicting time series are ARMA and ARIMA models.

6.2 Machine Learning Prediction

One important area of artificial intelligence (AI) is machine learning (ML) prediction [40] which uses statistical models to create predictive systems that can analyze data, spot trends and make predictions. Machine learning has transformed industries by allowing systems to adapt and optimize based on previous data. Machine learning is traditionally defined as a computer's capacity to learn from experience and enhance performance without explicit programming. From stock price prediction to fraud detection, machine learning algorithms are revolutionizing how businesses forecast trends and behaviors. As more data become accessible, these algorithms get better over time. They are trained on data to find patterns and forecast new, undiscovered data. The key to machine learning's success in a variety of fields, including network management, marketing, healthcare and finance, is its capacity to process massive datasets and produce predictions that are more and more accurate.

Selecting the appropriate machine learning method is contingent upon the nature of the task (e.g., regression or classification), the volume of data and the intended trade-off between performance and interpretability. For instance, linear regression may work well for small datasets with straightforward associations, while methods like random forests or neural networks may be needed for larger or more complicate datasets. The two main categories of machine learning algorithms are supervised and unsupervised learning. In order to predict results based on input data, supervised learning trains on labeled data for tasks like regression and classification. On the other hand, unsupervised learning, which is usually employed for dimensionality reduction and grouping, uses unlabeled data to reveal latent patterns.

7 ARMA, ARMAX, and ARIMA Models

7.1 ARMA

The Auto-Regressive Moving Average (ARMA) model [27–30] is a popular time series forecasting method that includes two components: auto-regressive (AR) and moving average (MA). Assuming that past trends or patterns impact future observations, the AR component uses historical values to forecast present values. This makes it possible for the model to identify any patterns or continuity in the data. In contrast, the MA component corrects future forecasts by taking into account prior forecast errors, which allows the model to adjust to potential random fluctuations and variances in the data. Because of this combination method, ARMA may be used for a number of forecasting applications, including estimating how many requests will arrive at each time step, which is represented by:

$$y(k) = \sum_{(k-1)T \leq t < kT} |R(t)|$$

As an example, assume ARMA is used to forecast temperature or precipitation in a particular area. Considering that weather patterns are consistent, the AR component forecasts the current weather based on historical rainfall or temperature data. For example, the AR component of the model might forecast that the temperature will keep rising if it has been rising consistently over the previous few days. Any inconsistencies or "misses" in earlier forecasts, such as an unforeseen cold front or an abrupt spike in rainfall, will be corrected by the MA component. For example, if an unpredicted rainstorm occurred on a day that was forecasted to be dry, the MA component incorporates this error to adjust the forecast for future predictions, making the model more responsive to sudden changes in weather patterns.

When applied to data that shows linear patterns or trends, ARMA's forecasting capability stems from its capacity to rely on historical values and error adjustments. The Mean Absolute Percentage Error (MAPE), which contrasts expected and observed values, is a frequently used performance indicator for ARMA. By providing information about the predictive dependability of ARMA, MAPE aids in evaluating how well it forecasts incoming data. For instance, ARMA usually produces more accurate forecasts than a more straightforward technique, such as the Last Value Prediction, when comparing the MAPE of the ARMA model for the number of incoming jobs, particularly when trends or variations are present in the data. As an example, taken from [30], Fig. 4 shows the MAPE of the number of incoming jobs for ARMA and Last Value predictions and it clearly demonstrates that ARMA yields more precise predictions compared to the simpler Last Value method, especially at the beginning of the experience, i.e. when there are few data.

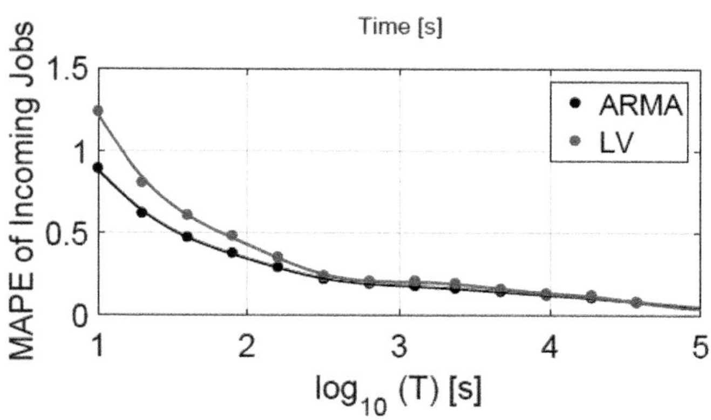

Fig. 4. MAPE of number of incoming jobs, for ARMA and Last Value predictions [30]

A qualitative comparison between ARMA and machine learning is presented in Table 1. It highlights that ARMA is simple, computationally inexpensive and works well with small amounts of data. However, it is limited to linear time series. Machine learning, on the other hand, is flexible and able to handle complex, nonlinear data. However, it requires significant resources.

Table 1. ARMA vs. Machine learning

Criteria	ARMA	Machine Learning
Data requirements	Requires large datasets for high accuracy	Works well with smaller datasets
Complexity	High	Simple
Computational Cost	High	Low
Prediction Power	Excellent for nonlinear, complex data	Best for linear, stationary time-series
Overfitting Risk	High	Low
Accuracy with Limited Data	May underperform with small datasets	More reliable with limited data
Adaptability	Highly	Limited

7.2 ARMAX

An expansion of the ARMA model that takes exogenous factors into account is the ARMAX model [27–30]. ARMAX adds exogenous variables, or outside variables, that are assessed separately from the system yet affect the forecasts, whereas ARMA models only use historical data and forecast mistakes. These exogenous variables, like arriving jobs when projecting exiting jobs, are utilized to increase the model's accuracy, particularly when the data being projected are greatly impacted by external sources.

There are multiple important steps in the ARMAX prediction process. Initially, at time step t, the model gathers data for the exogenous variable $x(t)$ (such as entering jobs as external influences), the dependent variable $y(t)$ (such as the number of outgoing jobs), and the error term $e(t)$, which takes into account departures from the predicted values. To anticipate the current value of $y(t)$, these gathered values are then entered into the ARMAX equation. To provide forecasts, the ARMAX model integrates the exogenous variables, the moving average component, and the auto-regressive component. In order to help the model adjust to new patterns and shifts in trends, it also employs a forgetting factor to progressively lessen the impact of earlier data.

Because it takes into account outside variables that affect the forecast, ARMAX typically exhibits higher prediction accuracy when compared to ARMA and Last Value models. In contrast to ARMA, which only uses historical values and errors, the ARMAX model modifies its forecasts of outgoing employment by taking incoming jobs into account as an exogenous signal. This produces more accurate results. The accuracy of these models is frequently evaluated using the Mean Absolute Percentage Error (MAPE), as shown in Fig. 5. The comparison indicates that ARMAX consistently performs better than both ARMA and the Last Value models, proving its capacity to lower the error margin and produce more accurate forecasts.

Unlike ARMA, the ARMAX equation, in Fig. 6, allows the integration of external influences since it incorporates an additional exogenous variable component.

7.3 ARIMA

As an expansion of the ARMA model too, the ARIMA model adds integration as a new element. Through the use of differencing techniques, which stabilize

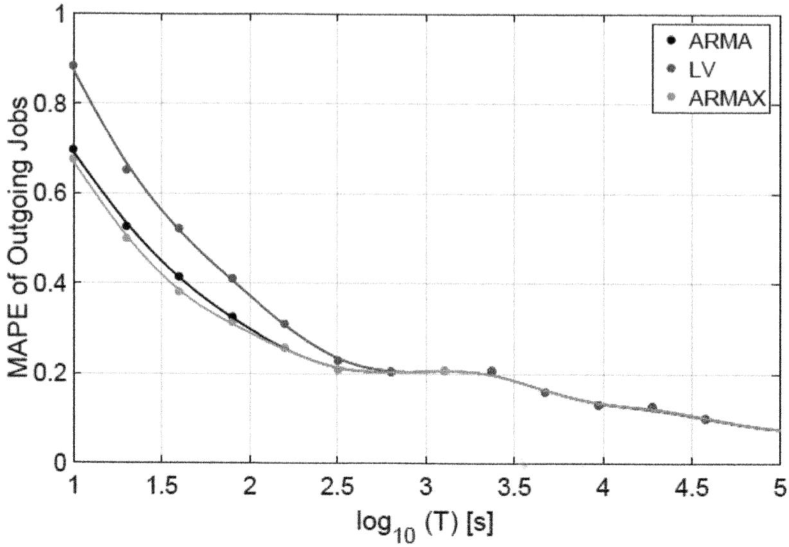

Fig. 5. MAPE of number of outgoing jobs using ARMA, ARMAX with incoming jobs as the exogenous signal, and Last Value models [30]

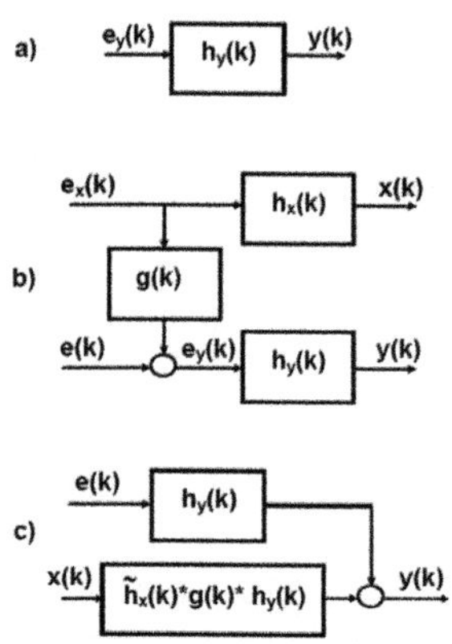

$$y(k+1) = h_y(k) * e_y(k+1) = h_y(0)e_y(k+1) + h_y^+(k) * e_y(k).$$

$$x(k) = h_x(k) * e_x(k)$$
$$e_y(k) = e(k) + g(k) * e_x(k)$$
$$y(k+1) = h_y(k) * (e(k+1) + g(k) * e_x(k+1))$$

$$\begin{aligned} y(k) =& a_1(k)y(k-1) + \cdots + a_n(k)y(k-na) \\ &+ b_1(k)x(k-1-nk) + \cdots + b_n(k)x(k-nk-nb) \\ &+ e(k) + c_1(k)e(k-1) + \cdots + c_n(k)e(k-nc), \end{aligned}$$

Fig. 6. ARMA & ARMAX [30]

trends over time and make the data suitable for forecasting, this integration allows the ARIMA model to handle non-stationary data. ARIMA's use to more complicated time series data is expanded since it can handle datasets with trends or seasonality, unlike ARMA, which implies stationarity.

The Auto-Regressive (AR), Integration (I) and Moving Average (MA) filters are the three main parts of the ARIMA model, as shown in the Fig. 7. Whereas the MA filter corrects forecasts using historical errors, the AR filter predicts future values based on historical values. The integration filter ensures stationarity and improves the accuracy of the model by differentiating the data to eliminate trends or seasonality.

Fig. 7. The ARIMA model

ARIMA has a number of benefits and drawbacks. Among its benefits, ARIMA just needs historical data, negating the need for extra explanatory variables and it is very good for short-term forecasting. It is adaptable across multiple domains and excels at modeling non-stationary data. ARIMA, however, is not intended for long-term forecasting and has trouble predicting turning points or sudden shifts in patterns. Its accuracy and efficiency may also be impacted by its computing demands and frequent reliance on arbitrary parameter choices.

The use of ARIMA in image analysis for cancer detection is a specific illustration of its effectiveness. As illustrated in the flow diagram presented in Fig. 8, the ARIMA model identifies anomalies in pixel patterns, assisting clinicians in detecting cancer at earlier stages. ARIMA improves patient outcomes and demonstrates the model's usefulness in real-world situations by assisting in early diagnosis through the analysis of non-stationary patterns in medical imaging data.

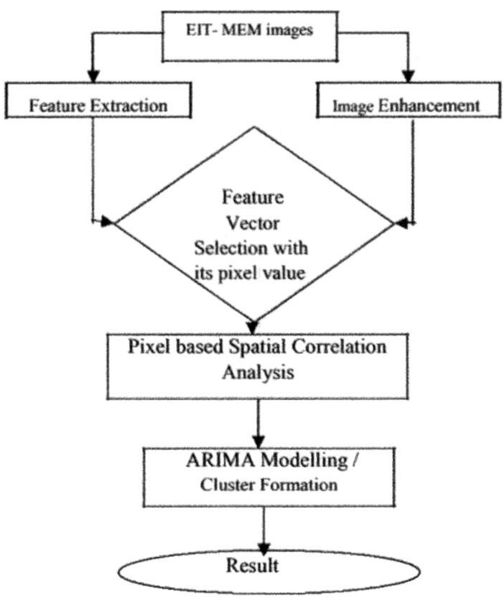

Fig. 8. Application of ARIMA in Identifying Anomalies in Medical Imaging [24]

8 Intents and Prediction

Auto-regressive models and statistical analysis are combined in the intent prediction and management process to guarantee effective use of available resources. Using these methods, the system finds important intent characteristics like CPU utilization, memory needs and bandwidth requirements, all of which are essential for comprehending and controlling network demands. Intents with comparable resource needs are grouped together when these characteristics are determined in order to expedite allocation and improve operational efficiency.

The technology predicts future intents based on past data and trends to guarantee flexibility and relevance. This proactive strategy guarantees that resources are prepared to meet future needs and makes it possible to anticipate them. Predictive models are also updated often to incorporate fresh data and increase accuracy, enabling ongoing development and flexibility in response to changing network circumstances and demands.

While machine learning techniques offer flexibility and advanced capabilities, they often face challenges when dealing with limited datasets, which can hinder their performance. In such cases, the ARMA family of models, including ARIMA and ARMAX, provides a more data-efficient solution. These models are particularly suited for scenarios where historical data are scarce, making them a reliable and practical choice for forecasting and managing intents in resource-constrained environments.

9 Conclusion

Intent languages are pivotal in enabling efficient communication and management in networking contexts. ARMA models provide simplicity and efficiency for predictive modeling in intent-based systems, which makes them perfect for small-scale and linear data. On the other hand, machine learning excels at managing intricate patterns and sizable datasets and offers flexibility and adaptability.

Our next step is to investigate how these predictive techniques might improve intent-driven systems, utilizing the accuracy of ARMA and the adaptability of ML to boost resource management and decision-making across a range of applications.

References

1. 3gpp. https://www.3gpp.org/
2. Generations, languages. https://www.encyclopedia.com/computing/news-wires-white-papers-and-books/generations-languages
3. Graph models. https://graph.build/resources/graph-models
4. IETF. https://www.ietf.org/
5. Intelligence brief: Does intent matter in network automation? https://www.gsmaintelligence.com/media-coverage/intelligence-brief-does-intent-matter-in-network-automation/

6. Natural language processing (NLP). https://www.sas.com/fr_ca/insights/analytics/what-is-natural-language-processing-nlp.html
7. tmforum. https://www.tmforum.org/
8. What is natural language processing (NLP)? https://www.ibm.com/topics/natural-language-processing
9. What is SDN? https://www.nomios.com/resources/software-defined-networking-sdn/
10. Zero-touch network and service management (ZSM); reference architecture (2019). chrome-extension://efaidnbmnnnibpcajpcglclefindmkaj/https://www.etsi.org/deliver/etsi_gs/ZSM/001_099/002/01.01.01_60/gs_ZSM002v010101p.pdf
11. Domain-specific language (2023). https://www.devx.com/terms/domain-specific-language/
12. Abhashkumar, A., Kang, J.M., Banerjee, S., Akella, A., Zhang, Y., Wu, W.: Supporting diverse dynamic intent-based policies using janus. In: Proceedings of the 13th International Conference on emerging Networking EXperiments and Technologies, pp. 296–309 (2017)
13. Alcock, P., Simms, B., Fantom, W., Rotsos, C., Race, N.: Improving intent correctness with automated testing. In: 2022 IEEE 8th International Conference on Network Softwarization (NetSoft), pp. 61–66. IEEE (2022)
14. Anderson, C.J., et al.: Netkat: semantic foundations for networks. ACM SIGPLAN Not. **49**(1), 113–126 (2014)
15. Angi, A., Sacco, A., Esposito, F., Marchetto, G., Clemm, A.: NLP4: an architecture for intent-driven data plane programmability. In: 2022 IEEE 8th International Conference on Network Softwarization (NetSoft), pp. 25–30. IEEE (2022)
16. Foster, N., et al.: Frenetic: a network programming language. ACM Sigplan Not. **46**(9), 279–291 (2011)
17. Gude, N., et al.: NOX: towards an operating system for networks. ACM SIGCOMM Comput. Commun. Rev. **38**(3), 105–110 (2008)
18. Jacobs, A.S., et al.: Hey, lumi! using natural language for {intent-based} network management. In: 2021 USENIX Annual Technical Conference (USENIX ATC 2021), pp. 625–639 (2021)
19. Jain, P.: Generations of programming language (2018). https://www.includehelp.com/basics/generations-of-programming-language.aspx
20. Kasneci, E., et al.: Chatgpt for good? On opportunities and challenges of large language models for education. Learn. Individ. Differ. **103**, 102274 (2023)
21. Kim, H., Reich, J., Gupta, A., Shahbaz, M., Feamster, N., Clark, R.: Kinetic: verifiable dynamic network control. In: 12th USENIX Symposium on Networked Systems Design and Implementation (NSDI 15), pp. 59–72 (2015)
22. Koroteev, M.V.: BERT: a review of applications in natural language processing and understanding. arXiv preprint arXiv:2103.11943 (2021)
23. Kozen, D.: Kleene algebra with tests. ACM Trans. Program. Lang. Syst. (TOPLAS) **19**(3), 427–443 (1997)
24. Kumar, N., Kumari, P., Ranjan, P., Vaish, A.: Arima model based breast cancer detection and classification through image processing. In: 2014 Students Conference on Engineering and Systems, pp. 1–5. IEEE (2014)
25. Liu, Y., et al.: Roberta: a robustly optimized BERT pretraining approach. arXiv preprint arXiv:1907.11692 (2019)
26. Martin, L., et al.: Camembert: a tasty French language model. arXiv preprint arXiv:1911.03894 (2019)

27. Milocco, R., Minet, P., Renault, E., Boumerdassi, S.: Evaluating the upper bound of energy cost saving by proactive data center management. IEEE Trans. Netw. Serv. Manage. **17**(3), 1527–1541 (2020)
28. Milocco, R., Minet, P., Renault, E., Boumerdassi, S.: Proactive data center management using predictive approaches. IEEE Access **8**, 161776–161786 (2020)
29. Milocco, R., Minet, P., Renault, É., Boumerdassi, S.: An evaluation method of optimal cost saving in a data center with proactive management. Manag. Data Center Netw. 105–127 (2021)
30. Milocco, R., Minety, P., Renault, E., Boumerdassi, S.: Cost reduction bounds of proactive management based on request prediction. In: 2019 International Conference on High Performance Computing & Simulation (HPCS), pp. 864–871. IEEE (2019)
31. Monsanto, C., Reich, J., Foster, N., Rexford, J., Walker, D.: Composing software defined networks. In: 10th USENIX Symposium on Networked Systems Design and Implementation (NSDI 2013), pp. 1–13 (2013)
32. Narayana, S., et al.: Language-directed hardware design for network performance monitoring. In: Proceedings of the Conference of the ACM Special Interest Group on Data Communication, pp. 85–98 (2017)
33. Nordquist, R.: What is a natural language? (2020). https://www.thoughtco.com/what-is-a-natural-language-1691422
34. Prakash, C., et al.: PGA: using graphs to express and automatically reconcile network policies. ACM SIGCOMM Comput. Commun. Rev. **45**(4), 29–42 (2015)
35. Nabilla, R.: What is a query: understanding database queries (2024). https://www.hostinger.com/tutorials/what-is-a-query
36. Ryzhyk, L., et al.: Correct by construction networks using stepwise refinement. In: 14th USENIX Symposium on Networked Systems Design and Implementation (NSDI 2017), pp. 683–698 (2017)
37. Sakor, A., Singh, K., Patel, A., Vidal, M.-E.: Falcon 2.0: an entity and relation linking tool over wikidata. In: Proceedings of the 29th ACM International Conference on Information & Knowledge Management, pp. 3141–3148 (2020)
38. Soulé, R., et al.: Merlin: a language for provisioning network resources. In: Proceedings of the 10th ACM International on Conference on emerging Networking Experiments and Technologies, pp. 213–226 (2014)
39. Sung, Y.W.E., Tie, X., Wong, S.H., Zeng, H.: Robotron: top-down network management at Facebook scale. In: Proceedings of the 2016 ACM SIGCOMM Conference, pp. 426–439 (2016)
40. VM Software House. 8 machine learning algorithms for predictions (2024). Accessed 06 Dec 2024
41. Wijnen, B.: (nemo) intent language. IETF95, Buenos Aires (2016)
42. Zhou, T.: Nemo language (2016). https://wiki.onosproject.org/display/ONOS/NEMO+Language

Design and Evaluation of a Lightweight SDN Controller for Integrated Road and Rail Networks

Dingyang Liu[1], Dereje Mechal Molla[1], Leo Mendiboure[2(✉)], Sassi Maaloul[2,3], Marion Berbineau[1], and Hakim Badis[4]

[1] COSYS-LEOST, Univ. Gustave Eiffel, 59650 Villeneuve d'Ascq, France
[2] COSYS-ERENA, Univ. Gustave Eiffel, 33067 Pessac, France
leo.mendiboure@univ-eiffel.fr
[3] JUNIA Engineering School, 59000 Lille, France
[4] LIGM, Univ. Gustave Eiffel, 77454 Marne La Vallée, France

Abstract. Implementing reliable communications is a major challenge for rail and road communications networks. Current V2X (Vehicle-To-Everything) technologies come up against major limitations. These include the existence of white zones and incomplete coverage by cellular networks (4G/5G). This implies the joint use of multiple access technologies at the level of each vehicle and a real-time and transparent transition between these radio access technologies. The definition of optimal handover mechanisms based on Artificial Intelligence (AI) tools is part of the answer to this problem. However, the definition of a high-performance network manager, guaranteeing low latencies, is also an important challenge. Therefore, this paper proposes a novel Software-Defined Network (SDN) controller designed to optimise dynamic handovers between wireless interfaces based on real-time network conditions. It also considers Machine Learning (ML) to determine performance parameter requirements. This approach improves the reliability, flexibility and adaptability of in-vehicle communication systems, helping to increase the efficiency of Cooperative Intelligent Transportation Systems.

Keywords: SDN Controller · Rail and Road · V2X Connectivity technologies · Machine Learning · Radio Interfaces Management

1 Introduction

Demand for reliable and efficient communication in vehicular networks grows because of the emergence of critical applications with important Quality of Service (QoS) requirements [1]: remote/tele-operated driving, automated vehicles management, platooning, etc. What's more, these services are increasingly tending to be multi-modal (vehicles that can operate on both rail and road), as demonstrated by the many current projects aimed at developing such systems, notably the FerroMobile and Flexy projects in France [2,3].

F. Hacène et al. (Eds.): MLN 2024, LNCS 15540, pp. 133–148, 2026.
https://doi.org/10.1007/978-3-032-00552-6_10

Different V2X communication technologies such as ITS-G5 and C-V2X [4,5] could be seen as a way to achieve the expected level of performance for multi-modal vehicular services. ITS-G5 provides low-latency, short-range communications and is suitable for safety-critical applications, while C-V2X offers better integration with existing cellular networks and wider coverage. However, both technologies have limitations, including limited network coverage, unstable performance at high speeds, difficulties with channel resource allocation, and security issues. Moreover, these technologies need more standardization, and seamless interoperability poses significant challenges in mixed rail and road networks [6].

Previous studies have proposed some methods to address these challenges, but there are still limitations. For example, [7] used the M/G/1 queueing model to analyze network connectivity, showing that proper base station numbers and transmission range can achieve desired connectivity. However, this costly solution seems unrealistic. Alternatively, another idea that has emerged is that the vehicle can draw on all the access technologies available by modifying the behaviour of its Radio interface as required [8]. Software Defined Networking (SDN), decoupling network control plane and data plane, is the technology that has been most widely considered for such an implementation. Indeed, it enables dynamic, software-based management of radio interfaces. However, work to date has been based solely on existing controllers (e.g. Ryu, ONOS [9]) not designed for high-mobility environments such as the vehicle, and therefore leading to significant delays that are too high for critical applications [10]. Furthermore, as highlighted in the related works, studies aimed at optimising these systems has not yet focused simultaneously on the ideas of mobility management, real-time interface management and taking account of multiple access technologies.

This study proposes a new framework for road and rail environments based on a novel SDN controller that ensures fast response and stable and reliable communication in highly dynamic environments. The SDN controller issues flow rules for interface switching based on the communication data received from the SDN enabled nodes. This aims at improving the reliability, scalability and efficiency of the on-board communication system. Both the operation, an implementation and an initial evaluation of the article are presented here. This work could represent a first step towards the definition of an SDN controller specifically adapted to the vehicular environment.

The rest of this paper is structured as follows: Sect. 2 reviews related work. Section 3 presents the network architecture and SDN integration. Section 4 covers implementation details, and Sect. 5 evaluates system performance. Finally, Sect. 6 concludes the paper and outlines future directions.

Table 1. COMPARISON OF RELATED WORK (Sorted by Similarity to Research)

References	Research Content	Innovation Points	SDN-Based Technology	Real-Time Dynamic Management	Multi-Network Integration	Adaptive Interface Selection
[11]	Proposes redundant architecture to improve autonomous vehicle safety	Introduces hardware redundancy and communication fault tolerance	×	×	✓	×
[12]	Proposes SDN-based architecture for managing sensor networks in hazardous environments	Introduces fault recovery and resource management for static networks	✓	×	×	×
[13]	Discusses cellular V2X challenges like network congestion and resource allocation	Proposes solutions to enhance V2X performance	×	×	✓	×
[14]	Proposes SDN framework for Li-Fi and Wi-Fi integration	Cross-layer optimization between Li-Fi and Wi-Fi	✓	×	✓	×
[15]	Proposes cluster-based radio resource management in dynamic vehicular networks	Introduces a vehicle matching algorithm and dynamic clustering for link stability	×	✓	×	✓
[16]	Compares SDN controllers in safety-critical vehicular applications, focusing on delay and throughput	Benchmarks SDN controllers for vehicular networks	✓	✓	×	✓
[17]	Studies SDN support for IEEE 802.11 and C-V2X in V2X communication	Proposes SDN framework to manage both standards	✓	✓	✓	×
[18]	Proposes hybrid DSRC and C-V2X routing in cluster-based networks	Optimizes communication for dense vehicular networks	✓	✓	✓	×
[19]	Focuses on improving handover management in 5G SDN-based V2X communication	Reduces handover latency using SDN, especially with X2 interface	✓	✓	✓	✓

2 Related Work

By focusing on work aimed at integrating SDN technology into a vehicular environment, we have been able to identify three main types of study:

- Focus on network robustness: [11–13] mainly focus on enhancing the robustness of the network through redundancy and fault-tolerant architectures. Their core lies in ensuring the system's reliability in the event of a fault or dangerous scenario. These projects involve anticipating breakdowns, white zones and the management of critical services. However, most of these studies aim at static or predictable network environments and fail to adequately address the needs of dynamic mobile networks;
- Focus on performance optimization: [14–16] propose various SDN-based network optimization methods (for example base station selection, resource reservation), with a particular focus on resource management and security-critical applications. While these studies propose effective performance optimization frameworks, they mainly focus on specific technologies or some static features of SDN and fail to cover the dynamic management of heterogeneous networks comprehensively;

- Focus on multi-radio acess technologies management: [17–19] have significantly addressed the challenges of V2X communications, especially in terms of multi-communication technology integration and handover management. However, these methods still need a complete solution for real-time adaptive interface selection, which is critical for communication stability in high-speed mobile scenarios.

This brief comparison, supplemented by a more detailed analysis of the environment in Table 1, highlights the shortcomings of the work proposed to date. The main objectives of this article are 1) to propose an SDN-based solution that takes advantage of the potential of this technology, and 2) to offer real-time management of network interfaces by considering different access networks and by being able to switch dynamically between different interfaces. To the authors knowledge, no work has yet been done to address these various elements. Consequently, the aim of this article is to fill this gap by proposing a solution adapted to the constraints of vehicular networks.

3 Considered Network Architecture

Our aim in this work is to design an architecture managing multiple radio access technologies (cellular, WiFi, Longe Range, and potentially other technologies) and the associated access points and interfaces thanks to an SDN-based approach. As shown in Fig. 1, such an architecture will be composed of two main elements: the Control Center and the network. The system presented in this section, in contrast to conventional SDN solutions, aims to integrate lightweight components designed to be seamlessly integrated into the vehicle architecture.

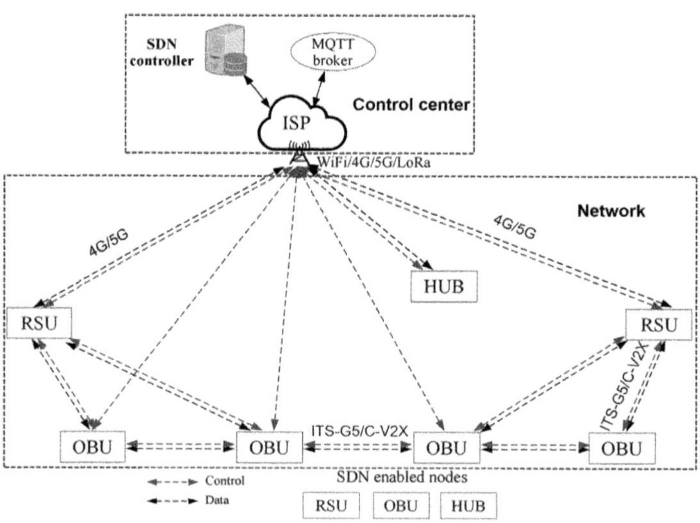

Fig. 1. SDN Architecture for Rail-Road Vehicles.

3.1 Control Center

The control center is the core component that implements the SDN function. It consists of an SDN controller (described in Sect. 4) and a Message Queuing Telemetry Transport (MQTT) agent [20]. The SDN controller centrally controls and dynamically manages network communication resources, optimizing routing, allocating spectrum, and managing wireless interface switching to ensure communication quality and connection reliability.

This controller connects with the entire network architecture through an Internet Service Provider (ISP). The ISP infrastructure mainly includes eNodeB/gNodeB (eNB/gNB) base stations and LoRaWAN Gateway, which provides 4G/5G and LoRa Radio Access Network (RAN) coverage for the system. The SDN controller communicates with network elements via the southbound interface in the SDN context. A MQTT protocol, which has a low bandwidth and efficient data transmission characteristics [20], is used to establish the southbound interface between the SDN controller and the SDN nodes.

3.2 Network

The network comprises multiple SDN-enabled nodes, such as Roadside Units (RSUs), Onboard Units (OBUs), and train stations (HUBs), managed by an SDN controller to form the overall communication architecture. These nodes are interconnected via various wireless technologies, including ITS-G5, C-V2X, 4G/5G, WiFi, and Long Range (LoRa), and support both single-hop and multi-hop communications, adapting to diverse traffic environments. Each node exchanges real-time data through ITS-G5 and C-V2X, and connects to the control center via 4G/5G and/or LoRa networks, enabling global monitoring and data transmission. In addition, LoRa can be used to exchange configuration information between the control center and SDN enabled nodes. This architecture forms a flexible multi-layer network supporting Vehicle to Vehicle (V2V), Vehicle to Infrastructure (V2I), and Vehicle to Network (V2N) communications, ensuring efficient and stable services.

4 Lightweight SDN Controller Implementation

After this quick introduction of the main components of the architecture (control center, network), this section will focus on the mechanisms implemented at SDN controller level to enable flow rules management and at SDN-enabled nodes to enable data transmission and collaboration between them.

4.1 Internal Architecture of SDN-Based Vehicular Networks

As shown in Fig. 2, the SDN controller and the SDN nodes (*e.g.*, OBUs, RSUs, HUBs) are the two key components of this experimental system. The node ensures real-time communication with the SDN controller by continuously sending status updates (such as position, speed, and QoS). It can also receive flow rules from the controller and use the switching manager to switch communication interfaces. This collaboration effectively meets the high reliability and efficiency requirements of V2X communication. Therefore, components need to be implemented at both levels to reproduce an SDN-based operation.

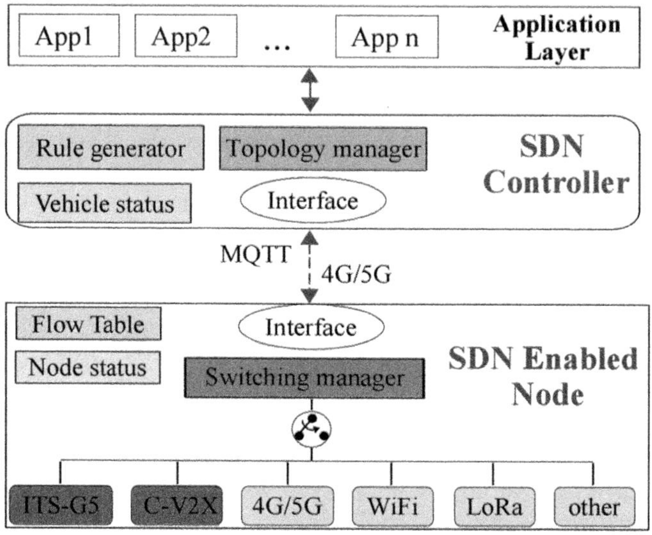

Fig. 2. Architecture of SDN-Based Vehicular Networks.

4.2 Flow Rule Generator

At the SDN controller, an essential element is the flow rule generator. Indeed, the flow rule generator in the SDN controller defines its core mechanism for issuing flow rules (routing rules formulation for SDN nodes). The structure of a flow rule is mainly divided into three parts, as shown in Fig. 3:

1. Match - what characterises the flow?
2. Action - what do we do with this flow?
3. Statistics - What details are available about implementing this flow?

Implementation details of each of these elements is provided below. The "Num" is mainly used for an SDN-enabled node to distinguish between disabled flow tables of the same "Command Type" and "OBU-ID".

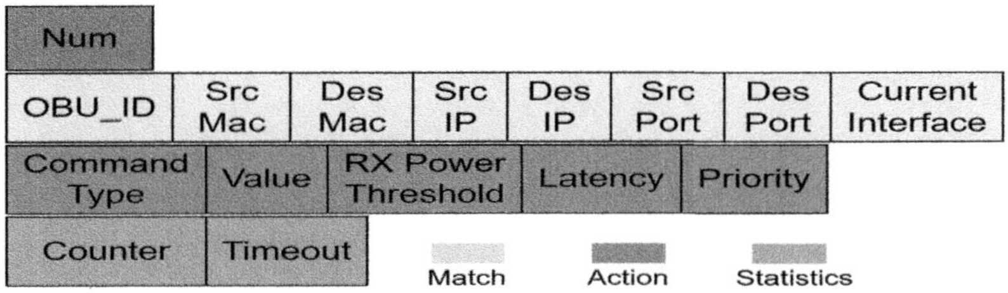

Fig. 3. Structure of a Flow Rule.

Match. The flow rule section is a prerequisite for executing actions. It defines the characteristics of data packets corresponding to the flow rule (e.g., Source/Destination MAC, IP, port, Interface, *etc.*), determining how to process the packet [10]. This is essential for enabling the SDN architecture to function.

Action. This part defines execution thresholds and priorities based on communication data and perform the corresponding action when conditions are met. Each SDN enabled node receives "Command Type", "Value", "RX Power Threshold", and "Latency" from the SDN controller if the values in the Match field matches with their identification. The Command Type could be switching radio interfaces, or forwarding to next node. The SDN enabled node performs the action defined in the Value field based on comparison of the real time value of RX Power and/or Latency with their threshold. At this level, our idea was to offer a multi-layer solution, as opposed to traditional SDN architectures which do not necessarily integrate information from the low ISO layers.

That is why "Latency" and "RX Power Threshold", common parameters for ITS-G5 and C-V2X, are used as execution conditions in the flow rules (see Fig. 3). If Latency or RX Power exceeds or is below the values set by the SDN controller (based on applications' requirement which could be redefined by a network administrator or based on Machine Learning - ML), the condition is marked as high priority, while the other is set as low priority. Defining these conditions based on ML approach is described below.

Thresholds are determined using Machine Learning Models (high priority) and residual control (low priority). A linear regression model predicts Latency (L) or Power (P) based on features like Received Signal Strength Indicator (RSSI), Channel Busy Rate (CBR), Packets Per Second (PPS), Channel Busy Percentage (CBP) and payload. With the near-linear relationship between features and target variables, linear regression is chosen for its simplicity, continuous variable efficiency, and suitability for real-time vehicular communications.

$$\hat{y} = w_1 x_1 + w_2 x_2 + \cdots + w_n x_n + b \tag{1}$$

where $\hat{y}$ is the predicted value, which could be latency $\hat{L}$ or power $\hat{P}$; $\mathbf{x} = [x_1, x_2, \ldots, x_n]$ represents the input feature vector which is depending on the current QoS; $w_1, w_2, \ldots, w_n$ are the learned weights, and b is the bias term.

The model is trained by minimizing the Mean Squared Error (MSE) to optimize the weights $\mathbf{w}$ and bias b:

$$L(\mathbf{w}, b) = \frac{1}{N} \sum_{i=1}^{N} (\hat{y}_i - y_i)^2 \tag{2}$$

where N is the number of samples, $\hat{y}_i$ is the prediction for the i-th sample, and y_i is the true value.

The model predicts latency or power thresholds based on input characteristics and uses these predictions as real-time system criteria. By minimizing the error between predicted and actual values during training, it accurately captures network changes and adjusts thresholds dynamically in real-time.

When the high priority threshold is triggered, the low priority threshold is dynamically adjusted based on the residual between the high priority's predicted and actual values. Let $V = \{L, P\}$ represent the variable for either latency (L) or power (P). The residual for V is calculated as:

$$e_V = |V_{\text{actual}} - \hat{V}| \tag{3}$$

The low priority threshold is adjusted as:

$$V_{\text{adjusted}} = \hat{V} - k_V \cdot e_V \tag{4}$$

Here, k_V is the adjustment factor for V, where V_{actual} represents the actual latency or power, and $\hat{V}$ is the predicted threshold from the machine learning model. The residual e_V represents the deviation between expected and actual values, with the low priority threshold adjusted by reducing $\hat{V}$ in proportion to the residual. The adjustment factor k_V determines the impact of this residual.

By calculating the residual for high priority conditions, the system adjusts the low priority threshold based on actual network performance, ensuring coordinated handling of high and low priorities under varying network conditions.

Statistics. The Statistic part includes Timeout and Counter, as depicted in Fig. 3, elements relating to the communication flow. Timeout represents the duration of the flow rules remaining active before expiring, while Counter tracks how often the user device applies the flow rules. These are the usual components of SDN architectures, but in theory they are based on an arbitrary definition of these elements. The idea here is to link them to the mobile nature of the nodes and the availability of the communication networks. In another words, a node need to be able to switch its radio interfaces while within the coverage range of the current active radio interface to maintain connectivity. Thus, the timeout varies in relation to the speed, direction and position of the nodes (vehicles). Consequently, the algorithm for the Timeout mechanism is defined as follows:

If the direction is towards the boundary (D = 1):

$$time\_to\_boundary = \frac{C - P}{S} \tag{5}$$

If the direction is away from the boundary (D = −1):

$$time\_to\_boundary = \frac{P}{S} \tag{6}$$

Here, S represents the device speed (in $meters/second$), P is the device's current position relative to the boundary, D indicates movement direction (1 for towards the boundary, −1 away from the boundary), and C is the device's maximum coverage range (in $meters$).

4.3 Switching Manager

As shown in Fig. 2, this component is part of the SDN enabled node, designed to connect with various network technologies such as ITS-G5, C-V2X, 4G/5G, WiFi, and LoRa. It is therefore a component that manages interfaces and is integrated at node level. Its purpose is to enable real-time switching. An implementation of such a switching manager based on the use of Software-Defined Radio (SDR) device is presented in Sect. 5.

4.4 Dynamic Adjustment of Data Sending Frequency

To prevent the SDN controller from making incorrect decisions, SDN enabled nodes should dynamically adjust their data transmission frequency based on current conditions. In the flow rules, high latency, low power, and short timeout values signal the need to increase the transmission frequency to ensure accurate decisions by the SDN controller. The specific formula is as follows:

Let $V = \{\text{latency}, \text{power}, \text{timeout}\}$ represent the variable for latency, power, or timeout. The weight for each parameter is calculated as:

$$V_{\text{weight}} = \begin{cases} \frac{\text{latency}-L_{\min}}{L_{\max}-L_{\min}} & \text{for latency,} \\ \frac{V_{\max}-V}{V_{\max}-V_{\min}} & \text{for power and timeout.} \end{cases} \tag{7}$$

$$combined\_weight = w_L \times latency\_weight + w_T \times timeout\_weight + w_P \times power\_weight \tag{8}$$

$$adjusted\_frequency = f_{\min} + (1 - combined\_weight) \times (f_{\max} - f_{\min}) \tag{9}$$

where most of the variables in Eq. 7, 8, and 9 are self explanatory. The weighted coefficients (w_L, w_T and w_P) adjust the importance of each factor, allowing the algorithm to compute results for real-time applications.

The various components identified as being necessary for the operation of our SDN architecture have been defined in this section. In the next section, we demonstrate the deployment of SDN controller.

5 First Deployment of the SDN Rail-Road Controller

In this section, we present the environment we have considered for an initial deployment and evaluation of the SDN controller we have proposed. For this demonstration, we have considered two OBUs to form a two-hop network infrastructure through two wireless technologies, our custom SDN controller, southbound interface through MQTT, and an ISP based on 4G/5G RAN. The detailed network components and their implementation is described below.

5.1 Network Components Selection and Implementation

A simple schematic of deployment components is shown in Fig. 4.

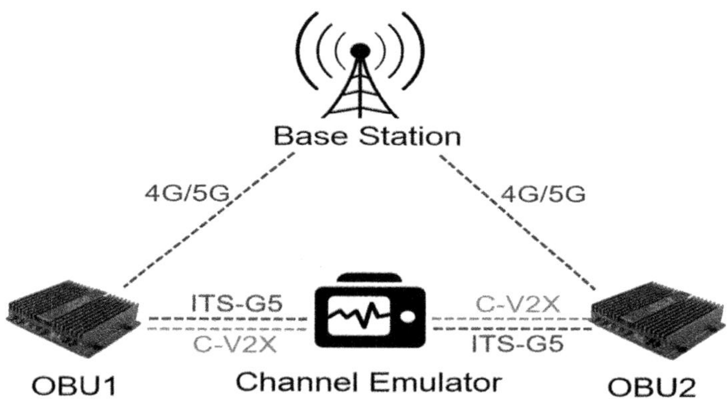

Fig. 4. Network and Channel Emulator.

Technical Choices for Implementing the Architecture. Cellular networks are widely used in vehicle communication systems for their high-speed, low-latency connections, essential for real-time data transmission [21]. For this reason, in this first experiment, a 4G/5G RAN is considered to enable communication between SDN controller and the SDN enabled nodes. The equipment (OBUs) are therefore connected to a base station via cellular communication. It is assumed here that it is capable of guaranteeing relatively reliable communications between the SDN controller and the SDN-enabled nodes (OBUs). Network selection takes place at the level of direct communications between OBUs. At this link, two radio access technologies are available (ITS-G5/C-V2X) and the system will aim to determine in real time which is the best technology. The variation in performance of these communication links is reproduced using a channel emulator. To sum up, our objective is to evaluate an initial real-time management system for multiple radio access interfaces.

Base Station and User Equipment Emulation. In this study, the base station consists of an eNB, a key component of the 4G LTE network, responsible for wireless communication with User Equipment (UE) and forwarding data to the core network. The eNB is implemented using an open-source RAN platform as presented in [22]. It uses a SDR platform based on a general purpose processor (Linux based workstation) and USRP B210 SDR board for flexible signal transmission. The Evolved Packet Core (EPC) handles packet routing, connection control, and network interface management, which is deployed on the same workstation as the eNB. All functions of the eNB and EPC are implemented using srsRAN 4G open-source RAN software [23]. The SDN enabled nodes act as the User Equipment in the 4G/5G terminology where each node is equipped with a SIM card module. With each component deployed and configured, we have a system that enables the implementation of an end-to-end 4G network and guarantees the communication between the control center (SDN controller) and the network elements (SDN enabled nodes).

Channel Emulator Deployment. The channel emulator in this study, built with GNU Radio, includes USRP sources and sinks, a dynamic channel model, and a frequency analysis module. The USRP components simulate signal transmission and reception, while the dynamic channel model replicates real-world effects like multipath propagation, fading, Doppler, and noise. As an initial test, an Extended Vehicular A (EVA) [5] channel model was designed with GNU Radio. This setup enables accurate V2X communication performance testing under various channel conditions in a controlled lab environment.

5.2 Global View of the Considered Platform

Figure 5 shows the platform that integrates the main components of the considered system and that was used for evaluating our new SDN controller:

- A workstation called SDN Controller Terminal in Fig. 5 that integrated both 1) the SDN controller that we developed, based on the components described in Sect. 3; 2) the channel emulator, required to reproduce varying channel conditions; and 3) the components required to run both 4G access and core network (srsRAN and Open5GS [24]);
- A laptop called SDN Enabled Nodes Terminal in Fig. 5 is used to access and configure the Linux based OBU boards that represent the SDN-enabled nodes, through serial/Ethernet interface. In this setup, an Ethernet interface is used to log-in to the OBU boards. The laptop is also used to display output results from each OBU board using separate terminal window;
- OBU-1 and OBU-2 represent the two SDN-enabled nodes used to form the network infrastructure, as shown in Fig. 4. They are based on Cohda Wireless MK6 OBU boards [25]. While ITS-G5 and C-V2X interfaces of the OBU are used to form the network between the two OBUs, cellular interface of the OBU is used to send vehicle status information to the SDN controller and receive flow rules from the SDN controller. A Global Navigation Satellite

System (GNSS) antenna is also used to obtain vehicle location information, and for time synchronization of the two OBU boards.
- USRPs are used in eNB and channel emulator, and their main purpose is signal reception and transmission and Downlink(DL) and Uplink(UL) communication. Since the two OBUs are connected to a single ISP (in this case one eNB), signal splitters (shown as SPs in Fig. 5), is used to divide one input signal to multiple output signals, allowing the same signal coming from the eNB to be used by multiple devices (OBUs). Indeed, the signal power after the splitter can be affected slightly (around 3 dB by the SPs used in our demonstrator). Similarly, other two splitters are used between the channel emulator and the two OBUs. The power degradation due to the splitter (*i.e.*, 3 dB value) is taken into account in our experimentation.
- The Switch is able to connect OBU-1 and OBU-2 to the laptop described above so that it helps to avoid using multiple laptops for each OBU.

Fig. 5. General framework of the experimental platform.

6 Time-Consuming Evaluation

Real-time and reliable data transfer is crucial for traffic safety and road efficiency in vehicular communication systems. As noted in [21], even slight delays can significantly impact system responsiveness and overall network performance. In this dynamic environment, both the management of in-vehicle communications and time-sensitive issues must be addressed. High vehicle speeds can cause delays, leading to missed critical decision windows and affecting performance. That is why this first experiment aimed at demonstrating that our SDN controller could guarantee real-time management of multiple radio access technologies. For the evaluation part, as with the earlier discussion on time impact in dynamic environments, we will focus on evaluating time consumption within this framework.

6.1 Defining Evaluation Metrics and Analyzing Results

This study will explore the time the SDN controller takes to send flow rules and the switching manager consumes per switch.

Testing 1 Time Taken to Send Flow Rules. The test measured the time taken to send flow rules in the EVA scenario at different Signal-to-Noise Ratio (SNR) levels (5 dB, 15 dB, 30 dB) and different vehicle speeds (10 km/h, 30 km/h, 50 km/h, and 70 km/h). The idea is to demonstrate that our solution, from the SDN controller perspective, remains functional under these variable conditions. The flow rule transmission times are presented in Table 2.

Table 2. Time taken to send flow rules

SNR Level	Speed Gap (km/h)	Time Taken (ms)	Distance (m) based on Speeds 30 km/h, 50 km/h, 70 km/h
5.0	10.0	24.596	0.205/0.342/0.478
5.0	30.0	24.272	0.202/0.337/0.472
5.0	50.0	24.582	0.205/0.341/0.478
5.0	70.0	23.819	0.198/0.331/0.463
15.0	10.0	24.164	0.201/0.336/0.470
15.0	30.0	27.427	0.228/0.380/0.532
15.0	50.0	24.098	0.200/0.335/0.469
15.0	70.0	23.691	0.197/0.329/0.461
30.0	10.0	23.857	0.199/0.332/0.465
30.0	30.0	22.970	0.191/0.320/0.449
30.0	50.0	22.958	0.191/0.320/0.448
30.0	70.0	25.227	0.210/0.351/0.491

As shown in Table 2 the time to send traffic rules ranges from 22.9 ms to 27.5 ms, depending on the SNR and speed differences. The distance covered at typical vehicle speeds is between 0.19 m and 0.54 m, a small range with minimal impact on receiving traffic rules within the communication coverage area.

Testing 2 Time Taken to Switch Interfaces. This test focuses on the average and stability of the time taken to switch between ITS-G5 and C-V2X to reflect possible impacts. The idea is to demonstrate that our solution, from the SDN-enabled devices perspective, remains functional under these variable conditions (RX power, latency). In order to simulate the performance under different communication conditions, it is based on the adjustment of the emulator parameters in the eNB configuration file. And, it is based on the previously formulated

ML algorithm that the internal value settings, parameter integration, and flow rule transmission are used to calculate the time consumed. The result for this testing is presented in Table 3.

Table 3. Time taken to switch interfaces

Switching Time	C-V2X to ITS-G5	ITS-G5 to C-V2X
Maximum (s)	10.60	11.62
Minimum (s)	2.01	2.22
Mean (s) excluding Max&Min	3.62	3.73
Median (s) excluding Max&Min	3.22	3.25
Standard Deviation (s) excluding Max&Min	1.63	1.76
Distance based on different speed	30.17/50.28/70.39	31.08/51.81/72.53

As given in Table 3, it takes an average of 3.62 s to 3.73 s to switch between different technologies. During this period, a vehicle travels around 30 m to 73 m. This means that the switching must be completed within 75 m to avoid communication interruption, which is generally acceptable for low speed vehicles since ITS-G5 and C-V2X have a coverage range of more than 300 m [17]. However, at high speeds, reducing the switching time is critical for maintaining communication continuity and ensuring safety [21]. The switching time depends on the software and hardware implementation of the wireless technologies within the MK6 OBU boards and implementation of our switching manager module. A future work focuses on optimization of this switching time. These preliminary assessments seem to show that our solution could be capable of guaranteeing the extremely short lead times required for critical services.

7 Conclusions and Future Work

This study defined and implemented a new SDN controller for wireless vehicular networking in railway and road environments. The solution enhances connection stability by dynamically managing the switching between communication interfaces based on real-time conditions, supporting cooperative intelligent transportation systems. Initial evaluations suggest that our solution could be capable of guaranteeing extremely short timescales for managing multiple radio interfaces (in the order of milliseconds) in mobile SDN nodes, and could therefore represent a major step forward compared with traditional SDN solutions that mainly focused on either a wired network or stationary wireless nodes.

In future work, it is possible to further optimize the ML algorithms used in this study, particularly in predicting latency and RX power thresholds, to improve the speed and accuracy of interface switching decisions. The optimization process will explore more advanced models to reduce the computational overhead while ensuring real-time performance. Reducing the interface switching time is also critical to maintaining communication stability, especially in high-speed environments. Additionally, future efforts will explore emerging technologies like 6G to extend coverage and enhance performance in areas with limited infrastructure. Furthermore, applying our SDN controller in a multi-hop scenario, and strengthening the security of SDN controllers against network attacks and data breaches will be another priority.

Acknowledgments. The work presented was performed in the framework of the French FLEXMOVE project supported by the ADEME agency within the FRANCE 2030 program and the European PODS4RAIL project GA 101121853.

References

1. Maaloul, S., Aniss, H., Berbineau, M., Mendiboure, L.: Evolution of use cases for intelligent transport systems. In: Cooperative Intelligent Transport Systems: Control and Management, p. 33 (2024)
2. Brahim, R., Collart-Dutilleul, S., Bon, P., Laharotte, P., Faouzi, N.: FERROMOBILE and security for low moment of traffic level crossing. In: International Conference on Risks and Security of Internet and Systems, pp. 35–50 (2023)
3. Jodeau, J., Absi, N., Chevrier, R., Feillet, D.: The rail-road Dial-a-Ride problem. Eur. J. Oper. Res. (2024)
4. ETSI Standard EN 302 663 V1.2.1: Intelligent Transport Systems: Access layer specification for Intelligent Transport Systems operating in the 5 GHz frequency band (2013)
5. 3GPP Standard TR 36.300 v14.3.0: Evolved Universal Terrestrial Radio Access (E-UTRA) and Evolved Universal Terrestrial Radio Access Network (E-UTRAN), Overall description; Stage 2 (Release 14) (2017)
6. Rammohan, A., et al.: Revolutionizing intelligent transportation systems with cellular vehicle-to-everything (C-V2X) technology: current trends, use cases, emerging technologies, standardization bodies, industry analytics and future directions. Veh. Commun. 100638 (2023)
7. Yousefi, S., Altman, E., El-Azouzi, R., Fathy, M.: Improving connectivity in vehicular ad hoc networks. Comput. Commun. **31**, 1653–1659 (2008)
8. Sylla, T., Mendiboure, L., Maaloul, S., Aniss, H., Chalouf, M., Delbruel, S.: Multiconnectivity for 5G networks and beyond: a survey. Sensors **22**, 7591 (2022)
9. Sheikh, M., Halder, M., Kabir, S., Miah, M., Khatun, S.: SDN-based approach to evaluate the best controller: internal controller NOX and external controllers POX, ONOS, RYU. Global J. Comput. Sci. Technol. **19**, 21–32 (2019)
10. McKeown, N., et al.: OpenFlow: enabling innovation in campus networks. Comput. Commun. Rev. **38**, 69–74 (2008)
11. Boubakri, A., Gammar, S.: A new architecture of autonomous vehicles: redundant architecture to improve operational safety. Int. J. Robot. Control Syst. **1**, 355–368 (2021)

12. Weinstein, S.B.: SDN control in a combined sensor, information retrieval, and communications network for dangerous environments. In: Proceedings of IEEE International Conference on Microwaves, Antennas, Communications and Electronic Systems (2017)
13. Gyawali, S., Xu, S., Qian, Y., Hu, R.Q.: Challenges and solutions for cellular based V2X communications. IEEE Commun. Surv. Tutor. **23**(1), 222–255 (2021)
14. Alshaer, H., Haas, H.: SDN-enabled Li-Fi/Wi-Fi wireless medium access technologies integration framework. In: IEEE Conference on Standards for Communications and Networking (2016)
15. Huang, J., Cui, H., Chen, C.: Cluster-based radio resource management in dynamic vehicular networks. IEEE Access **10**, 43562–43570 (2022)
16. Smida, K., Tounsi, H., Frikha, M., Song, Y.-Q.: Efficient SDN controller for safety applications in SDN-based vehicular networks: POX, Floodlight, ONOS or OpenDaylight? In: Proceedings of IEEE Eighth International Conference on Communications and Networking (ComNet) (2020)
17. Nurkahfi, G., et al.: On SDN to support the IEEE 802.11 and C-V2X based vehicular communications. IEEE Access (2023)
18. Zhang, Y., Zhang, J.: Design and optimization of cluster-based DSRC and C-V2X hybrid routing. Appl. Sci. **12**(13), 6782 (2022)
19. Rizvi, H., Akram, J.: Handover management in 5G software defined network based V2X communication. In: Proceedings of the International Conference on Open Source Systems and Technologies (ICOSST) (2018)
20. Manandhar, S.: MQTT based communication in IoT. M.Sc. thesis, Tampere University of Technology (2017)
21. Khattak, Z.H., Rios-Torres, J., Fontaine, M.: Impact of communications delay on safety and stability of connected and automated vehicle platoons: empirical evidence from experimental data. IEEE Access (2023)
22. Molla, D.M., Badis, H., George, L., Berbineau, M.: Coverage extension of software defined radio platforms for 3GPP 4G, 5G radio access networks. In: 2021 13th IFIP Wireless and Mobile Networking Conference (WMNC), Montreal, QC, Canada, pp. 55–62 (2021). https://doi.org/10.23919/WMNC53478.2021.9619041
23. srsRAN Project: Open Source RAN. https://www.srslte.com/. Accessed 10 Sept 2024
24. Mamushiane, L., Lysko, A., Kobo, H., Mwangama, J.: Deploying a stable 5G SA testbed using srsRAN and Open5GS: UE integration and troubleshooting towards network slicing. In: 2023 International Conference On Artificial Intelligence, Big Data, Computing And Data Communication Systems (icABCD), pp. 1–10 (2023)
25. Cohda Wireless, MK6. https://www.cohdawireless.com/solutions/mk6/. Accessed 17 July 2024

PiPS: An Effective Strategy and Approach for Privacy in Public Surveillance

Sajid Ahmed[1,2](✉) and Noriaki Yoshiura[2]

[1] Shah Abdul Latif University, Khairpur, Pakistan
ahmed.s.285@ms.saitama-u.ac.jp, sajid.ghanghro@salu.edu.pk
[2] Saitama University, Saitama, Japan
yoshiura@fmx.ics.saitama-u.ac.jp

Abstract. Public surveillance systems are known to serve extremely important purposes in many cities around the world. However, they create serious doubts regarding privacy concerns. This paper resolves that by using advanced face detection and recognition algorithms to protect individual privacy. Our approach relies on YOLOv8(You Only Look Once version 8) is a state-of-the-art object detection model that improves previous iterations of the YOLO series for face detection and has been trained on both face detection and WIDERFACE datasets (The WiderFace dataset is a large-scale benchmark designed for face detection tasks). In order to have only the face of the individual under tracking, we employ a face matching algorithm for its accurate recognition. Other faces in the surveillance video are blurred by the average filter to preserve the individual privacy. Thus, public security alongside the need for protection of personal privacy is a delicate balancing act that offers a robust solution for privacy in systems of public surveillance. Our experimental results demonstrated that our approach makes privacy possible without compromising the behavior of the system as far as public space monitoring and security are concerned.

Keywords: Privacy Preservation · Face matching · Public Surveillance

1 Introduction

Surveillance systems publicly installed have formed a backbone of urban security strategy on the global level. In turn, these will give law enforcement and other security agencies powerful tools for real-time monitoring of criminal activities, preventing crime, and maintaining public order. Equipped with increasingly sophisticated technologies such as facial recognition, these systems can identify people with remarkable accuracy, trace the movement of targets across locations, and analyze large-scale video data in real-time. However, this is also a double-edged sword, where the same capabilities that make those systems so effective raise considerable concerns regarding privacy. The continuous capturing, analysis, and storing of personal data, especially facial data, may lead to unauthorized profiling, identity theft, and the invasion of personal privacy, creating ethical and legal challenges well beyond that cannot be ignored.

F. Hacène et al. (Eds.): MLN 2024, LNCS 15540, pp. 149–169, 2026.
https://doi.org/10.1007/978-3-032-00552-6_11

Therefore, public security and individual privacy constitute the balance of tension that is a crucial issue in surveillance system deployment. The more pervasive these systems become, the greater the potential for misuse or abuse of individual privacy. Therefore, it is increased public debate and pressure with respect to demands for measures to ensure protection of privacy. Traditional methods of privacy protection in surveillance footage include data anonymization, secure multi-party computation, and homomorphic encryption [1]. These usually comprise methods for anonymizing data, which either mask or modify identifiable information in such a way that unauthorized access is not possible. While secure multiparty computation and homomorphic encryption enable the processing of data in a secure manner without exposure of the raw data itself. However, these approaches usually involve significant drawbacks. For instance, anonymization can reduce the effectiveness of surveillance by obscuring the critical details; many secure computation methods similarly incur heavy computation overhead, which is impractical in real-time applications over large-scale surveillance networks.

These challenges have been overcome by recent advancements in machine learning and computer vision to open new ways of developing privacy-preserving surveillance systems [2]. Among the many developments in these fields, one of the most promising object detection breakthroughs is the YOLO family of models, a game-changer in the field of object detection because of its unprecedented speed and accuracy. YOLOv8 is purposed for real-time processing of visual data, which would make them ideal to apply in a wide range of applications requiring both efficiency and precision, such as face detection and recognition in surveillance systems [3]. The model attains better detection capability with reduced computational demands; hence, this framework is considered a great stride toward the development of effective and scalable solutions in surveillance.

This paper presents a novel system for privacy-preserving surveillance based on the powerful YOLOv8 model for accurate and efficient face detection. Our approach incorporates a robust face matching algorithm to guarantee that in the whole sequence, only the faces of targeted individuals will remain visible, while the rest of the detected faces are blurred using an average filter. This effectively anonymizes the non-target individuals through a blurring technique, answering concerns about privacy without reducing the capabilities of the surveillance system to monitor or maintain safe public areas.

To further enhance privacy protection, our system is designed with stringent access controls that ensure personal data is not shared with unauthorized entities. Only designated users, such as law enforcement agencies or municipal authorities, are granted access to the retained footage and facial recognition data. Additionally, individuals have the option to request access to their own data by submitting a single verification picture of themselves. This request is subject to verification processes to ensure authenticity, allowing individuals to exercise control over their personal information. By implementing this framework, we aim to build public trust in surveillance systems, assuring individuals that their data is safeguarded against misuse and that privacy remains a fundamental consideration in the deployment of our technology. This commitment to ethical data management not only fortifies the integrity of our surveillance approach but also aligns with evolving legal standards and societal expectations regarding privacy in the digital age.

Through extensive experimental validation, we verify that our approach not only retains high accuracy in both detection and recognition but also considerably alleviates computational overhead owing to the privacy-preserving methods. This study will therefore fall within the current development curve of privacy-preserving surveillance technologies. It provides a pragmatic and effective approach to individual privacy protection in public spaces. It further brings our method into a robust framing for the future of the surveillance system in an urban environment by balancing dual imperatives of public security with those of personal privacy.

We organize the rest of this paper as follows. In Sect. 2, we review related works on privacy preserving surveillance systems and face recognition techniques. We describe the proposed methodology and system architecture in Sect. 3, the experimental setup and results in Sect. 4, and discuss the results, their implications, and directions for future research in Sect. 5.

2 Literature Review

A The rapid development and deployment of the surveillance systems, especially for public areas, raise significant concerns regarding the preservation of privacy. The increasing deployment of surveillance systems by governments and organizations introduces tension between the goal of maintaining public safety while protecting the privacy of individuals [4]. This literature review discusses recent advances in methods concerning privacy preservation; particularly on face detection, face recognition, and mechanisms for preserving privacy.

Different questions have raised on the uses of surveillance technologies regarding how these technologies relate to and affect marginalized communities. In this regard, Turner Lee and Chin [5] review the trends in the historical misuse of technologies related to surveillance and impacts on communities. Such technologies often reinforce societal biases, leading to unfair treatment and further surveillance of already vulnerable groups [6]. The authors call for robust federal privacy protections and regulatory oversight that would prevent discriminatory practices. Their work underlines the need for surveillance systems that respect civil liberties and are transparently designed so benefits derived from the same do not come at the cost of eroding individual freedoms.

Further, Königs [7] elaborates on the idea of "control privacy" versus "access privacy." He draws a fine line between how data collection and access are to be understood in any surveillance system. Control privacy will mean the right of any individual to determine who shall have access to his or her personal data. While access privacy will imply the mechanism for how data is accessed and used [8]. The critical difference between two concepts provides a basis on which ethical surveillance systems that provide the right to privacy for individuals should be developed.

Thence forms the ground for most contemporary-day surveillance technologies. Technological progress is occurring in such a manner that recent development strains a high capability for face detection and recognition systems that is capable of enhancing accuracy, speed, as well as preservation of privacy [9].

The fields that have been transformed by deep learning include face detection and recognition. Mahmoud and Mengash proposed [10] a hybrid approach combined with normalized RGB and YCbCr color spaces to detect the presence of occluded faces in videos taken by surveillance cameras [11]. The proposed system effectively segments skin areas and detects contours of head and shoulder, allowing an appreciable detection rate of concealed faces [12]. The practical applicability of the technique in real-time applications validates the technique for application in present-day systems, particularly where the individual may attempt to evade detection.

Kanagamalliga et al. extended this further by incorporating the use of Haar Cascade and CNNs within video surveillance systems [13]. The main novelties of their system involve edge computing and cloud-based solutions for scalability and adaptability. The system's ability to handle various lighting conditions and occlusions makes it particularly suited for real-world applications, where environmental factors often pose significant challenges.

Li et al. [14] contribute to the face identification imperative in a dynamic atmosphere by proposing a multi-task discriminative learning framework. It is implemented with a ConvNet model combined with a 3D mean face model and leverages the applications of facial key points along with 3D transformation parameters to enhance accuracy. The method in this paper is effective under "wild conditions," where variations in pose, light, and occlusions add more complications to face detection and recognition [15].

Hybrid models appeared to meet the demand for both processing efficiency and scalability in systems designed for face recognition. In this regard, Lei et al. [16] proposed RFR-DLVT-a hybrid system that combines deep learning with visual tracking. In a consideration to optimize the efficiency of the processing with minimal loss of recognition accuracy, video sequences are divided into Reference Frames (RF) and Non-Reference Frames (NRF). Applications will be useful in situations that require real-time monitoring, such as flowing public transportation hubs or large-scale events [17].

Wang et al. [18] came up with a feature subtraction-based method that was trainable to extend the domain of privacy-preserving face recognition. This minimizes visual feature representations while preserving identity information necessary for individual privacy protection. In other words, the method generates visually uninformative residues from original face images, hence guaranteeing that individual privacy remains safeguarded without necessarily compromising the accuracy of recognition systems [19]. This technique represents one major step toward balancing the dual objectives at stake: security and, simultaneously, privacy.

Advanced surveillance technologies being integrated into public spaces do call for the development of comprehensive regulatory frameworks. Turner Lee and Chin [20] recommend increased federal privacy protection and clearly spelled-out limits on the application of surveillance technologies. Their work evidences the need for regulatory frameworks that can forestall the misuse of surveillance data and ensure deployment in a manner that does not violate individual rights [21].

Apart from regulatory measures, the use of synthetic data during the training process of a face recognition model will offer improved levels of privacy [22]. Wang et al. established that effective training of the models was possible without access to personal biometric data by training the datasets with synthetic datasets. This approach also carries along with it the advantage of minimizing risks related to data breaches and unauthorized access to sensitive information [23].

Privacy refers to an individual's right or access to their information and how, when, and by whom their information is being collected, used or shared. It means protecting all sensitive information equal to the person's identity, dealings or communications, from being abused or accessed by other people [24]. Privacy enables people to exercise control over their personal lives and reduce the extent to which their activities are prone to unwelcomed observation or encroachment.

2.1 Privacy in Public Surveillance Systems

As public surveillance systems become more and more numerous, so does concern about them due to the fact that they have raised serious issues regarding privacy. These systems are intended to observe communal areas and the use of surveillance technologies, which includes face recognition systems, can result to the implementation of surveillance of non-suspect or non-target individuals [25]. This entails the likelihood of revealing sensitive private information about individuals that include their names, whereabouts and the activities that they participate in, which they would seek to keep confidential.

The paradox stems from the fact that these systems have two functions public safety and personal privacy. Traditional surveillance systems operate by capturing and storing vast amounts of video footage, including detailed facial information of all individuals in view. This introduces the potential for misuse, either by insiders accessing the data for non-legitimate reasons or through external data breaches, which can expose private, sensitive information [26].

Privacy preservation refers to the methods and techniques used to protect individuals' personal information while ensuring that the data remains usable for legitimate purposes. In the context of face recognition and video surveillance, privacy preservation involves techniques that prevent unauthorized access or misuse of sensitive information such as facial identities [27]. Several approaches have been developed to balance the need for privacy with the functional use of data in public surveillance systems. Table 1 shows various privacy preservation techniques.

Table 1. Privacy Preservation Techniques

S. No.	Technique	Description	Key Methods	Effectiveness
1	Obfuscation and Anonymization [4]	Conceals identities by transforming faces into a less identifiable form, allowing detection but preventing unauthorized facial recognition	Face blurring, Masking, Adversarial Obfuscation	Highly effective for protecting identities while maintaining detection functionality

(continued)

Table 1. (*continued*)

S. No.	Technique	Description	Key Methods	Effectiveness
2	Controllable Privacy Protection (CPP) [10]	Uses generative adversarial networks (GANs) or deep learning to adjust privacy levels, modifying faces to obscure identity while retaining certain attributes	GANs, Deep Learning Models, YOLO-based Architectures	Ensures adjustable privacy while preserving essential features like gender or expressions
3	Invertible Mask Networks (IMN) [20]	A reversible form of privacy preservation where the original face can be recovered by authorized users	Reversible Masking	Effective for scenarios requiring reversible access to the original data for authorized users
4	Blurred and Masked Faces [1]	Widely used in video surveillance to obscure facial details of non-target individuals, allowing privacy for bystanders while retaining the ability to retrieve data	Blurring, Masking, Average Filters	Effective in protecting non-relevant individuals' privacy, with the option of authorized data retrieval
5	Homomorphic Encryption and SMC [19]	Allows computations on encrypted data without revealing sensitive information, using multi-party computation (SMC) to distribute tasks across entities	Fully Homomorphic Encryption, Secure Multi-Party Computation	Provides strong encryption and privacy protection during data processing but requires high computational resources
6	Differential Privacy [23]	Introduces noise into datasets to anonymize individuals while allowing aggregate analysis, ensuring face images cannot easily be traced back to a specific person	Noise Addition, Perturbation Algorithms	Effective for large datasets where privacy needs to be preserved without affecting the aggregate analysis

(*continued*)

Table 1. *(continued)*

S. No.	Technique	Description	Key Methods	Effectiveness
7	Selective Detection and Recognition [20]	Focuses on identifying target individuals while anonymizing or ignoring non-target individuals in the frame, preserving privacy	YOLO-based Detection Algorithms	Effective in ensuring only relevant information is captured, preserving privacy for non-target individuals
8	Data Encryption and Access Control [8]	Encrypts captured data, with access to sensitive information restricted to authorized individuals, ensuring privacy from data collection to storage	Data Encryption, Access Control Policies	Highly effective for multi-layered privacy protection, from data collection to retrieval by authorized users
9	Low-Resolution Images [27]	Reduces the resolution of images to limit the amount of detailed personal information captured, maintaining privacy for non-relevant individuals	Lowering Image Resolution	Effective for privacy in surveillance while maintaining general utility of the footage
10	Loss Function Modifications for Privacy [28]	Optimizes loss functions in face detection algorithms to balance detection accuracy and privacy preservation, minimizing exposure of unnecessary details	Loss Function Optimization, YOLO-based Approaches	Effective in reducing unnecessary data exposure while maintaining detection accuracy
11	AI-Driven Privacy Masks [16]	Dynamically applies privacy masks in real-time using AI algorithms to obfuscate personal features while ensuring system functionality for face detection and recognition	AI-Driven Privacy Masking, Real-time Masking	Ensures real-time privacy protection without affecting face detection accuracy

(continued)

Table 1. (*continued*)

S. No.	Technique	Description	Key Methods	Effectiveness
12	Siamese Neural Networks for Face Matching [18]	Uses deep siamese neural networks to match faces, only retaining matched identities for further processing while discarding non-matching data	Siamese Neural Networks	Effective for selective data retention, ensuring privacy by discarding irrelevant data
13	Preprocessing and Data Augmentation [14]	Modifies input data through preprocessing techniques like lighting, angle adjustment, and facial feature transformation, enhancing privacy while retaining recognition	Preprocessing Techniques, Data Augmentation	Effective in reducing sensitive information while enabling accurate recognition, though it requires pre-analysis adjustments

3 Proposed System

The proposed work focuses on enhancing public surveillance systems by integrating advanced face detection and recognition techniques while prioritizing individual privacy. Utilizing the YOLOv8 model, our approach achieves accurate face detection trained on relevant datasets, such as Face detection and WIDERFACE. To ensure that only the target individual's face is visible, we implement a precise face matching algorithm. Additionally, to protect the privacy of all other detected faces, we apply an average filter to blur them in the surveillance footage. This method effectively balances the critical need for security in public spaces with the essential requirement to safeguard personal privacy, demonstrating its efficacy through rigorous experimental results (Fig. 1).

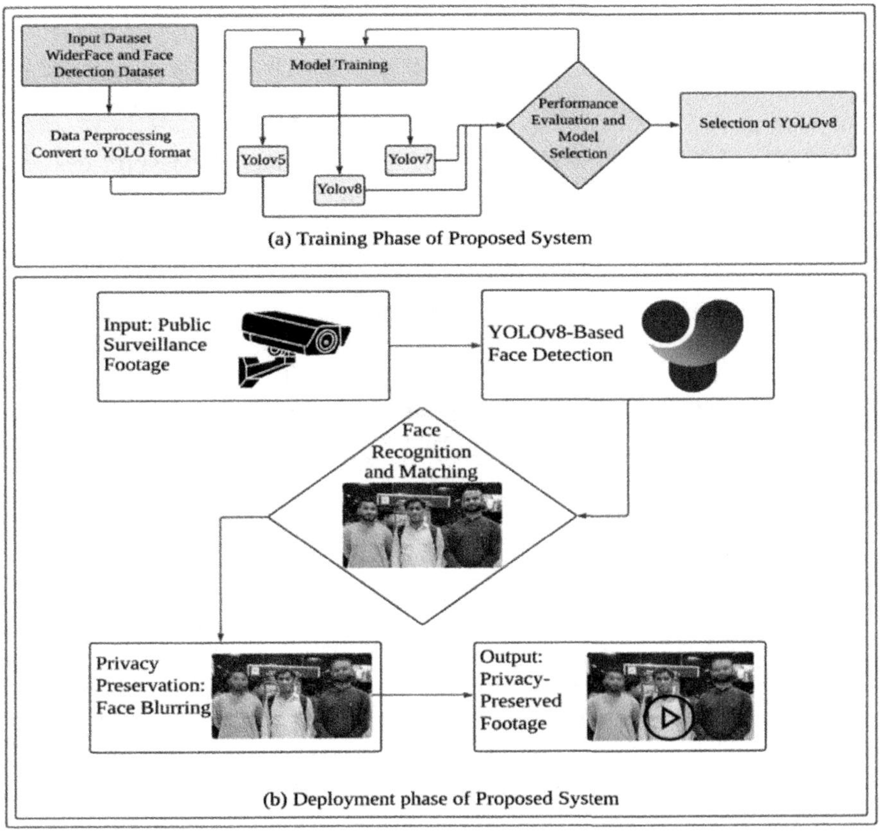

Fig. 1. Proposed System

3.1 Training Phase

3.1.1 Input Dataset (WiderFace and Face Detection Dataset)

First of all, there is a need to determine the input datasets that are going to be used. Main training data will rely on "WiderFace" and "Face Detection Dataset". Both these varied datasets contain different scenarios which include variation in lighting, orientations, and background to comprehensively train face detection.

3.1.2 Data Preprocessing (Convert into YOLO Format)

Then, the raw datasets will be preprocessed in YOLO format; that means resizing images, normalizing pixel values, and annotating the bounding boxes such that these fit the requirements of the YOLO model. The YOLO format usually would require image files with corresponding text files containing object class IDs, bounding box coordinates, and image dimensions.

3.1.3 Model Training

The pre-processed data then goes into model training. Here, three models are being trained: YOLOv5, YOLOv7, and YOLOv8. Each of these models undergoes multiple training epochs i.e., 10,50,100,150,200, and 300 learning the dataset for precise face detection. Training consists of adjusting model weights through backpropagation and optimization algorithms with the purpose of minimizing the error of detections.

3.1.4 Model Selection and Performance Evaluation

The models are evaluated, after training, on various metrics such as accuracy, precision, recall, F1 score, and mAP for the extent of each model's face detection performance for YOLOv5, YOLOv7, and YOLOv8. The purpose of a performance evaluation would be to make sure only the most accurate and proficient model is selected for further use. Further evaluation metrics are defined in result section.

3.1.5 Selection of YOLOv8

This flowchart depicts that, through performance evaluation, the selected model will be YOLOv8; it outperformed others by a big margin in accuracy, precision, and overall performance in face detection. Being a later version, it reaps more advanced features and improvements, thus making this version apt for this application.

The following flowchart summarizes the end-to-end process of training, starting from dataset preparation to training multiple versions of YOLO, selecting the best-performance model, which is YOLOv8, for the most accurate face detection tasks. This helps to pick only the best in model selection for real-world usage.

3.2 Deployment Phase

3.2.1 Input: Public Surveillance Footage

The process begins with input, which is the footage retrieved from public surveillance cameras; they would generally be located at spots in public view, such as streets, parks, airports, shopping malls, and/or transportation terminals. Most of the locations are indeed covered with continuous cameras, or much data containing faces of people passing through them. The data from surveillance cameras can be used for security monitoring, crowd behavior analysis, etc. Containing sensitive personal information about the individuals, this data and hence shall be dealt with carefully regarding privacy.

3.2.2 YOLOv8-Based Face Detection

It takes a captured surveillance footage and passed it through a deep learning-based face detection model called YOLOv8 model. YOLOv8: You Only Look Once version 8 is among the most recent and technologically advanced models of the YOLO series. It's known for real-time object detection at very high accuracy and efficiency. YOLOv8 will process the frames in the video where it is able to detect multiple faces in each frame. The model divides each frame into a grid and then outputs bounding boxes around the faces each with a confidence score indicating how much it thinks that a given box contains a face. YOLOv8 has the advantage in terms of speed and accuracy, which shall make it quite ideal for real-time video feed processing, where there is an urgent need for quick detection.

3.2.3 Face Recognition and Matching

When the YOLOv8 model detects faces within a footage, the next process is the recognition and matching of such faces. What this simply means is checking through a database for an identification of who the faces are.

Face Recognition: This is the process of analyzing facial features that include the general outline of the face, eyes, nose, mouth, and all the other distinguishing features that are used to identify a person. It may make use of pre-trained models or any other database to match faces against profiles of known individuals.

Matching Process: When it involves a database of known persons, the system tries to match the detected faces with entries in that database. This step is quite important in security applications where specific identification of an individual-like a suspect or a person of interest-is required.

The face matching function compares the facial encodings between an input image and a cropped face. This process can be mathematically described as follows.

Let:
E_{input} be the face encoding of the input image
$E_{cropped}$ be the face encoding of the cropped face region

The face matching function is represented by the equation:

$$R = compare(E_{input}, E_{cropped})$$

where:

$$R \in \{0, 1\}$$

- R = 1 if the faces match (True), $R = 0$ if the faces do not match (False).
- The comparison is made by calculating the distance d between the encodings and checking if it is less than a predefined threshold value:

$$R = \begin{cases} 1, \ if\ d(Einput, Ecropped) < threshold \\ 0, \ otherwise \end{cases} \tag{1}$$

where,

- $d(E_{input}, E_{cropped})$ is the distance (typically using Euclidean distance or cosine similarity) between the input and cropped face encodings.
- Threshold is the threshold value that determines if two faces are considered a match.

3.2.4 Privacy Preserving: Face Blurring

In essence, it is a way of blurring faces captured in video footage to keep the privacy of the persons captured. Considering the privacy laws, this is an important step to maintain ethical values while handling data from surveillance that involves the general public. YOLOv8 outputs in bounding boxes are used to identify the position of faces, and then a blurring filter is applied on that area. This is used to obscure the facial features so that the faces cannot be recognizable anymore.

Some methods to blur would involve the use of Gaussian blur, pixelation, or mosaic. The aim should be such that the resolution and clear view of the face are lost and it may no longer be recognizable. This ensures that while the video capture is useful for monitoring, it does not capture the fine details that would reveal an individual's identity or enable tracking.

The 2D average filter computes the average value of pixels in a local n × n neighborhood around a given pixel. Mathematically, the equation for the 2D average filter is:

$$\frac{1}{n^2} \sum_{k=-\frac{n}{2}}^{\frac{n}{2}} \sum_{l=-\frac{n}{2}}^{\frac{n}{2}} I(i+k, j+l) \tag{2}$$

Explanation:
Let:

- $I(i+k, j+l)$ represents the intensity (or pixel value) at position $(i+k, j+l)$ in the image.
- $iandj$ represent the current pixel's coordinates.
- n represents the size of the averaging kernel, typically a square filter such as 3×3, 5×5, etc.
- $\sum_{k=-\frac{n}{2}}^{\frac{n}{2}} \sum_{l=-\frac{n}{2}}^{\frac{n}{2}}$ sums over all pixel values within the $n \times n$ neighborhood centered at (i, j).
- $\frac{1}{n^2}$ is the normalization factor that ensures the sum is divided by the total number of pixels in the neighborhood (i.e., n^2 for an $n \times n$ kernel).

The result of this operation is the average intensity of the pixels within the $n \times n$ neighborhood, which replaces the original pixel value at position (i, j).

If the system does not find a match, or if privacy is to be maintained, then the process goes further for maintaining identity.

3.2.5 Output: Privacy-Preserved Footage

The final output would be a modified version of the original footage in which all of the detected faces have been blurred. Overall context and activity captured by cameras would still be preserved; however, these would not expose identities. Applications of Output: a) The processed footage can be shared with law enforcement, security personnel, or other stakeholders without risking exposure to personal identities. This privacy-preserved video can be used for analytics, such as crowd movements or determining public behavior, without violations of personal privacy rights.

The equation describing the process of blurring all faces except the target individual's face in video footage can be represented as:

$$F_{\text{output}}(i,j) = \begin{cases} F_{\text{original}}(i,j) & \text{if}(i,j) \in \text{Target's Face} \\ Blur\left(F_{\text{original}}(i,j)\right) & if\,(i,j) \in \text{Other Faces} \end{cases} \tag{3}$$

Explanation:

- $F_{\text{output}}(i,j)$ is the processed video frame at pixel location (i,j).
- $F_{\text{original}}(i,j)$ is the original video frame at pixel location (i,j).
- If the pixel (i,j) belongs to the target individual's face, the pixel value remains unchanged.
- If the pixel (i,j) belongs to another face, a blur function Blur() is applied to the pixel values to obscure the face.

Equation (3) formalizes the idea that only the target individual's face remains visible, while other faces are blurred to maintain privacy

4 Results and Experimental Setup

4.1 Accuracy

Accuracy measures the overall correctness of the model's predictions by indicating the proportion of true results (both true positives and true negatives) among the total number of predictions. It is defined as

$$\text{Accuracy} = \frac{\boldsymbol{TP} + \boldsymbol{TN}}{\boldsymbol{TP} + \boldsymbol{TN} + \boldsymbol{FP} + \boldsymbol{FN}}$$

where *TP is True Positives, TN is True Negatives, FP is False Positives, and FN is False Negatives.*

4.2 Precision

Precision indicates how many of the predicted positive instances were actually positive, measuring the model's exactness. It is calculated as

$$\text{Precision} = \frac{\boldsymbol{TP}}{\boldsymbol{TP} + \boldsymbol{FP}}$$

4.3 Recall (Sensitivity)

Recall measures how many of the actual positive instances were correctly identified by the model, assessing the model's completeness. It is given by:

$$\text{Recall} = \frac{\boldsymbol{TP}}{\boldsymbol{TP} + \boldsymbol{FN}}$$

A higher recall means fewer false negative errors.

4.4 F1 Score

The F1 score is the harmonic mean of precision and recall, providing a single metric that balances both. It is useful when you want to find a balance between precision and recall. The formula for the F1 score is

$$F_1 \text{ Score} = 2 \times \frac{\text{Precision} \times \text{Recall}}{\text{Precision} + \text{Recall}}$$

The F1 score is most informative when you need to account for both precision and recall, especially in cases where there is an uneven class distribution

4.5 Performance of YOLOv5 over Variation of Epochs (Face Detection Dataset)

The YOLOv5 model for face detection demonstrated consistent improvements across various training epochs, with precision, recall, and F1 score serving as key performance metrics. Initially, at 10 epochs, the model achieved an F1 score of 0.85 with an AUC of 0.884. Performance steadily improved at 50 epochs, reaching an F1 score of 0.86 and an AUC of 0.909. The model's optimal performance was observed between 100 and 200 epochs, where the F1 score stabilized at 0.87, and the AUC reached up to 0.914. By 300 epochs, the model effectively converged, maintaining the F1 score at 0.87 with a consistent AUC of 0.913, indicating that further training beyond this point provided minimal additional benefit.

4.6 Performance of YOLOv7 over Variation of Epochs (Face Detection Dataset)

The YOLOv7 model for face detection showed consistent improvements over different training epochs. At 10 epochs, it achieved an F1 score of 0.85 with an AUC of 0.882. By 50 epochs, the F1 score increased to 0.86, and the AUC improved to 0.908. At 100 epochs, the model sustained an F1 score of 0.86, with the AUC reaching 0.911. Optimal performance was seen around 150 epochs, where the F1 score was 0.87, and the AUC peaked at 0.914. The model effectively converged by 300 epochs, maintaining an F1 score of 0.86 and an AUC of 0.913. Although YOLOv7 exhibited steady improvements, it was slightly less optimal compared to YOLOv8 in this study.

4.7 Performance of YOLOv8 over Variation of Epochs (Face Detection Dataset)

The YOLOv8 model demonstrated significant improvements across different training epochs and consistently outperformed YOLOv5 in terms of speed, precision, and detection accuracy. At 10 epochs, it achieved an F1 score of 0.85. By 50 epochs, the F1 score improved to 0.86, and the AUC reached 0.911. The model maintained strong performance at 100 epochs, with a similar F1 score of 0.86. Optimal performance was observed at 150 epochs, where the F1 score reached 0.87. At 200 epochs, the model sustained an F1 score of 0.87 with an AUC of 0.913, and by 300 epochs, the model had effectively converged, demonstrating its superior learning capability over YOLOv5 and YOLOv7. This analysis highlights that YOLOv8 consistently excelled in face detection accuracy and overall performance as training progressed.

With YOLOv8 achieving a maximum accuracy of 0.94 compared to YOLOv5's 0.91 at 300 epochs and YOLOv7's 0.93 at 300 epochs, it demonstrates that YOLOv8 was more adept at learning and refining the task of face detection. Both models maintained a balanced F1 score between 0.87–0.88, indicating stable precision and recall, but YOLOv8's superior precision at higher epochs shows its enhanced ability to correctly identify faces with fewer false positives. This improvement is especially evident in more extensive training, where YOLOv8 maintained higher precision and accuracy, demonstrating its robustness in handling complex detection scenarios.

4.8 Performance of YOLOv5 over Variation of Epochs (WiderFace Dataset)

The YOLOv5 model for face detection showed continuous improvement across various training epochs, with key performance metrics demonstrating its learning progression. Starting at 10 epochs, it achieved an accuracy of 75.59% with an F1 score of 0.76. By 50 epochs, the accuracy increased to 79.29% with an F1 score of 0.79. At 100 epochs, the model improved further, achieving 81.92% accuracy and an F1 score of 0.82. The trend continued at 150 epochs with 83.24% accuracy and an F1 score of 0.84. By 200 epochs, the model reached 85.21% accuracy with an F1 score of 0.85. Finally, at 300 epochs, the model attained its peak performance with an accuracy of 86.38% and an F1 score of 0.87, indicating effective convergence and enhanced face detection capability with prolonged training.

4.9 Performance of YOLOv7 over Variation of Epochs (WiderFace Dataset)

The YOLOv7 model for face detection on the WiderFace dataset showed consistent improvements across various training epochs. Starting at 10 epochs, it achieved an F1 score of 0.74 with an AUC of 0.793. By 50 epochs, the F1 score increased to 0.77, with an AUC of 0.812. At 100 epochs, the F1 score rose to 0.79, and the AUC improved to 0.825. Further training at 150 epochs led to an F1 score of 0.81 with an AUC of 0.835. By 200 epochs, the F1 score reached 0.82 with an AUC of 0.841, and at 300 epochs, the model achieved its highest performance with an F1 score of 0.83 and an AUC of 0.847. These results indicate that YOLOv7 effectively learned from the dataset and reached convergence, showing steady improvements over time.

4.10 Performance of YOLOv8 over Variation of Epochs (WiderFace Dataset)

The YOLOv8 model demonstrated progressive improvements in face detection accuracy on the WiderFace dataset over different training epochs. Starting at 10 epochs, it achieved an accuracy of 70% with an F1 score of 0.75. By 50 epochs, the accuracy increased to 73%, and the F1 score reached 0.78. At 100 epochs, the model's accuracy improved to 75%, with an F1 score of 0.80. Training for 150 epochs led to further enhancement, with an accuracy of 76% and an F1 score of 0.82. At 200 epochs, the model achieved an accuracy of 77% and its F1 score peaked at 0.83. Finally, at 300 epochs, YOLOv8 reached its optimal performance, achieving an accuracy of 78% and an F1 score of 0.84, indicating effective learning and convergence. These results show that YOLOv8 consistently improved its performance throughout the training process.

Considering the results, YOLOv8 offers more consistent improvement across epochs and achieves competitive results. Therefore, if a balance between high precision and adaptability is required, YOLOv8 is recommended for face detection tasks, especially when dealing with complex datasets where model adaptability is crucial. Table 2 shows results of different versions of Yolo on various epochs.

4.11 Training Graphs Pictures

See Figs. 2 and 3.

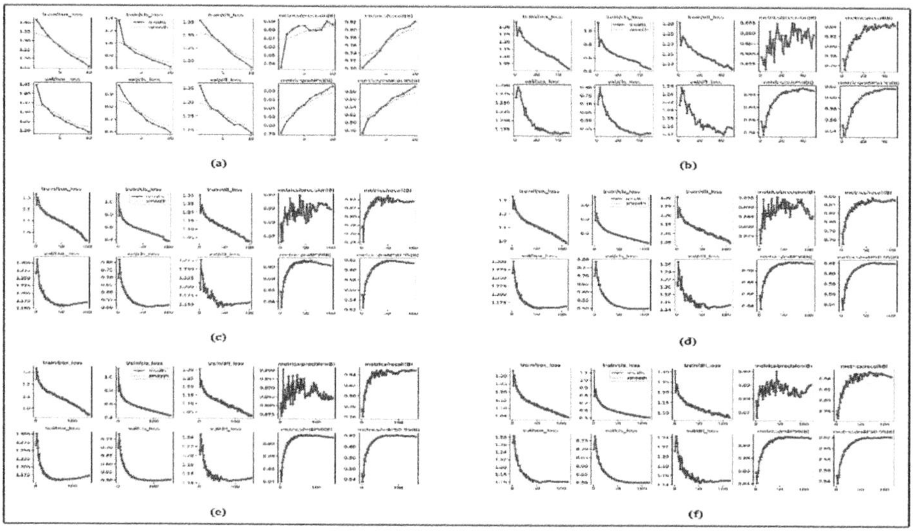

Fig. 2. Yolov8 on Face Detection Dataset

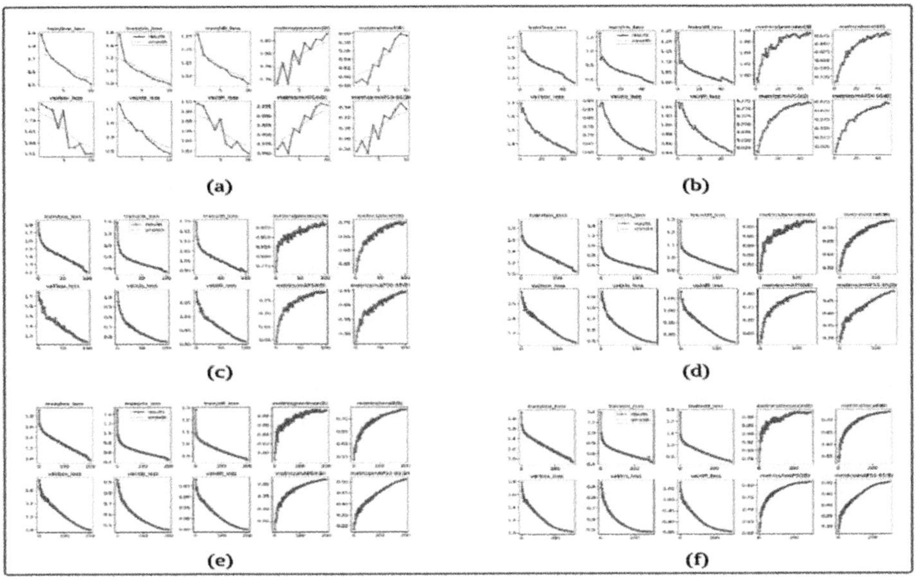

Fig. 3. WiderFace on yolov8

Table 2. Results Table

Dataset	Model	Epochs	Accuracy	Precision	Recall	F1 Score
Face Detection	YOLOv5	10	0.85	0.86	0.85	0.86
		50	0.85	0.88	0.87	0.87
		100	0.87	0.86	0.88	0.87
		150	0.85	0.89	0.87	0.88
		200	0.89	0.85	0.89	0.87
		300	0.91	0.86	0.89	0.87
	YOLOv7	10	0.86	0.85	0.85	0.85
		50	0.88	0.87	0.85	0.86
		100	0.9	0.87	0.85	0.86
		150	0.91	0.88	0.86	0.87
		200	0.92	0.84	0.88	0.86
		300	0.93	0.85	0.88	0.86
	YOLOv8	10	0.87	0.86	0.86	0.86
		50	0.89	0.88	0.86	0.87
		100	0.91	0.88	0.86	0.87
		150	0.92	0.89	0.87	0.88
		200	0.93	0.85	0.89	0.87

(*continued*)

Table 2. (*continued*)

Dataset	Model	Epochs	Accuracy	Precision	Recall	F1 Score
		300	0.94	0.86	0.89	0.87
WiderFace	YOLOv5	10	0.68	0.82	0.58	0.69
		50	0.76	0.82	0.65	0.76
		100	0.75	0.87	0.7	0.77
		150	0.78	0.88	0.75	0.79
		200	0.8	0.88	0.77	0.8
		300	0.81	0.89	0.78	0.8
	YOLOv7	10	0.69	0.84	0.64	0.74
		50	0.72	0.85	0.66	0.77
		100	0.74	0.87	0.68	0.79
		150	0.75	0.88	0.69	0.81
		200	0.76	0.89	0.71	0.82
		300	0.77	0.9	0.72	0.83
	YOLOv8	10	0.7	0.85	0.65	0.75
		50	0.73	0.86	0.67	0.78
		100	0.75	0.88	0.69	0.8
		150	0.76	0.89	0.7	0.82
		200	0.77	0.9	0.72	0.83
		300	0.78	0.91	0.73	0.84

4.12 Privacy Preservenetaion resutls

These images showcase the application of privacy preservation techniques using YOLOv8 for face detection, followed by the implementation of an average filter. The average filter effectively blurs the detected facial features, ensuring the privacy of individuals while maintaining the overall context of the scene. The result is a visually coherent image that balances privacy and concerns the need for clear visual representation. The pictures in Figure 4 are of the author, while the other images have been included with the consent of the individuals depicted.

Fig. 4. Privacy Preservation Image

5 Conclusion and Future Work

In this paper, we introduce PiPS, a novel solution that balances the trade-off between preserving privacy and public security in surveillance systems. Our approach leverages an advanced face detection capability from the YOLOv8 model along with a robust face matching algorithm that ensures only the faces of targeted subjects remain visible while non-target faces are anonymized by an average blurring filter. This approach provides a critical solution to the problem of privacy invasion in public surveillance without affecting the capability of the system in keeping public space monitored and secured. Experimental results have shown that PiPS can ensure real-time privacy protection with negligible degradation in detection accuracy and computational efficiency. Our solution is scalable and adaptable to different contexts, ranging from small-scale environments to extensive urban monitoring networks. This work falls squarely within the emergent area of privacy-preserving surveillance and provides a useful real framework for future systems that need to balance public safety imperatives with individual privacy concerns. Future work for PiPS includes integrating advanced anonymization techniques like GANs, improving recognition in multi-object and low-light environments, incorporating privacy technologies such as homomorphic encryption, and testing real-world performance under diverse conditions to enhance its overall effectiveness.

Acknowledgments. I want to convey my heartfelt thanks to my Supervisor Noriaki Yoshiura for his invaluable guidance and support during the research journey. Special thanks to Saitama University, Japan for providing the necessary financial support for this study. Thanks to my family and friends for their unwavering encouragement and understanding.

References

1. Anusudha, K.: Real time face recognition system based on YOLO and InsightFace. Multimedia tools and applications **83**(11), 31893–31910 (2024)
2. Fang, Y., Zhang, Y.: Face recognition approach for smart Internet of Things in home security system. Journal of optics **53**(2), 1203–1209 (2024)
3. Chen, L., et al.: Two-stage dual-resolution face network for cross-resolution face recognition in surveillance systems. The Visual Computer **40**(8), 5545–5556 (2024)
4. Chang, C.-Y., et al.: Design and implementation of a real-time face recognition system based on artificial intelligence techniques. Multimedia Systems **30**(2), 114 (2024)
5. Kanagamalliga, S., et al.: Advancements in real-time face recognition algorithms for enhanced smart video surveillance. Procedia Comp. Sci. **230**, 486–492 (2023)
6. Delussu, R., Putzu, L., Fumera, G.: Synthetic Data for Video Surveillance Applications of Computer Vision: A Review. Int. J. Comp. Vision, 1–37 (2024)
7. Aung, H., Bobkov, A.V., Tun, N.L.: Face detection in real time live video using yolo algorithm based on Vgg16 convolutional neural network. In: 2021 International conference on industrial engineering, applications and manufacturing (ICIEAM). IEEE (2021)
8. Karpagam, M., et al.: A novel face recognition model for fighting against human trafficking in surveillance videos and rescuing victims. Soft Computing **27**(18), 13165–13180 (2023)
9. Bentafat, E., Rathore, M.M., Bakiras, S.: Towards real-time privacy-preserving video surveillance. Computer Communications **180**, 97–108 (2021)
10. Diwan, T., Anirudh, G., Tembhurne, J.V.: Object detection using YOLO: Challenges, architectural successors, datasets and applications. Multimedia Tools and Applications **82**(6), 9243–9275 (2023)
11. Maitlo, N., et al.: Color Recognition in Challenging Lighting Environments: CNN Approach. In: 2024 IEEE 9th International Conference for Convergence in Technology (I2CT). IEEE (2024)
12. Khan, S.S., et al.: MTCNN++: A CNN-based face detection algorithm inspired by MTCNN. The Visual Computer **40**(2), 899–917 (2024)
13. Gaikwad, B., Karmakar, A.: Real-time distributed video analytics for privacy-aware person search. Comp. Vis. Image Underst. **234**, 103749 (2023)
14. Chen, W., et al.: YOLO-face: a real-time face detector. The Visual Computer **37**, 805–813 (2021)
15. van Daalen, O.: The right to encryption: Privacy as preventing unlawful access. Comp. Law Secu. Rev. **49**, 105804 (2023)
16. De Coninck, S., Leroux, S., Simoens, P.: Privacy aware person detection in surveillance data. arXiv preprint arXiv:2110.15171 (2021)
17. Meden, B., et al.: Face deidentification with controllable privacy protection. Image and Vision Computing **134**, 104678 (2023)
18. Gokulakrishnan, S., et al.: An optimized facial recognition model for identifying criminal activities using deep learning strategy. Int. J. Info. Technol. **15**(7), 3907–3921 (2023)
19. Qian, Y.: Face recognition technology for video surveillance integrated with particle swarm optimization algorithm. Int. J. Intel. Netw. **5**, 145–153 (2024)

20. Hu, W., et al.: Privacy-preserving face recognition method based on extensible feature extraction. J. Visual Comm. Image Represent. **100**, 104140 (2024)
21. Imran, A., et al.: FaceEngine: a tracking-based framework for real-time face recognition in video surveillance system. SN Comp. Sci. **5**(5), 609 (2024)
22. Tao, Y., et al.: Privacy-preserving outsourcing scheme of face recognition based on locally linear embedding. Computers & Security 103931 (2024)
23. Tomar, V., Kumar, N., Srivastava, A.R.: Single sample face recognition using deep learning: a survey. Artif. Intel. Rev. **56**(Suppl 1), 1063–1111 (2023)
24. Mostafa, S.A., et al.: A YOLO-based deep learning model for Real-Time face mask detection via drone surveillance in public spaces. Info. Sci. 120865 (2024)
25. Menon, S., et al.: Custom Face Recognition Using YOLO. V3. In: 2021 3rd International Conference on Signal Processing and Communication (ICPSC). IEEE (2021)
26. PK, R., et al.: Fraud detection and prevention by face recognition with and without mask for banking application. Multimedia Tools and Appl. 1–24 (2024)
27. Singh, A., et al.: Automation of surveillance systems using deep learning and facial recognition. Int. J. Sys. Assur. Eng. Manage. **14**(Suppl 1), 236–245 (2023)

A Comprehensive Review of Deep Learning Approaches for Tomato Leaf Diseases Detection and Classification in Smart Agriculture

Oussama Nabil(✉) and Cherkaoui Leghris

LMCSA (Laboratory of Mathematics, Computer Sciences and Applications), Faculty of Sciences and Techniques, Hassan II University of Casablanca, Casablanca, Morocco
oussama.nabil-etu@univh2c.ma, cherkaoui.leghris@fstm.ac.ma

Abstract. This review aims to provide an in-depth evaluation of various deep learning approaches used for the detection and classification of tomato leaf diseases, a critical challenge in smart agriculture. By examining a wide range of methods, the study seeks to identify the most effective deep learning models in terms of accuracy, speed, and robustness in real-world scenarios. Special emphasis is placed on both qualitative and quantitative performance analysis, including metrics such as precision, recall, and F1-score, to assess the effectiveness and reliability of the techniques. Image segmentation, including Convolutional Neural Networks (CNN) and Deep Neural Networks (DNN), plays a pivotal role in isolating and identifying diseased areas on leaves. Furthermore, the classification aspect is assessed to determine which methods offer the best differentiation between disease types. The review also highlights the technical challenges associated with deploying these models in diverse agricultural settings, such as variations in lighting, image capture angles, and image quality. The ultimate goal is to guide future research and technological advancements towards more efficient solutions for early disease detection, reducing agricultural losses, and enhancing large-scale tomato crop management.

Keywords: Deep Learning · Tomato Leaf Diseases · Smart Agriculture · Image Segmentation · Disease Classification · Agricultural Technology · Crop Management · Early Disease Detection

1 Introduction

Modern agriculture faces a major challenge: the early detection and management of crop diseases, particularly in tomatoes, which are a vital crop globally. The rapid spread of tomato leaf diseases can lead to significant agricultural losses, affecting not only production but also crop quality. In response to this issue, the use of advanced technologies, such as deep learning and smart agriculture systems, has emerged as a promising solution.

F. Hacène et al. (Eds.): MLN 2024, LNCS 15540, pp. 170–179, 2026.
https://doi.org/10.1007/978-3-032-00552-6_12

Deep learning approaches, in particular, stand out for their ability to analyze large amounts of images and automate the disease detection process. Among these techniques, Convolutional Neural Networks (CNN) and Deep Neural Networks (DNN) play a key role in image segmentation and disease classification from leaf images. Image segmentation allows for the isolation of affected areas, while classification helps differentiate between disease types, thereby improving diagnostic accuracy.

However, applying these models in real agricultural environments still poses technical challenges. Variations in lighting conditions, image capture angles, and differences in image quality can affect model performance. It is therefore essential to understand which techniques are the most robust and adapted to these constraints to maximize their effectiveness in the field. This study provides a review of deep learning methods used for detecting tomato leaf diseases, with a focus on segmentation and classification techniques. It aims to identify the most effective models and analyze the challenges and opportunities they present for improving large-scale crop management. The qualitative and quantitative analysis of model performance is essential for a comprehensive evaluation, including metrics such as precision, recall, and F1-score.

2 State of the Art

The classification of plant diseases is essential in the context of smart agriculture, especially for crops like tomatoes. Various techniques have been developed for effective diagnosis and management of these diseases. For example, Kai Tian [1] investigated segmentation methods using the K-means algorithm, which helps identify areas affected by diseases like Tomato Yellow Leaf Curl Virus (TYLCV) and Tomato Mosaic Virus (ToMV). Applied to a dataset of tomato leaf images, this method uses adaptive parameters, such as Lab color spaces, to adjust for variations in brightness and contrast, improving accuracy. Other clustering techniques, such as hierarchical clustering and Fuzzy C-means, have also been used to detect lesions from diseases like Tomato Blight, enhancing segmentation precision by grouping data points based on similarities.

Convolutional Neural Networks (CNNs) have proven to be powerful tools for disease classification and segmentation. Naresh K. Trivedi [8] developed a CNN model trained on a dataset of 3,000 tomato leaf images covering nine disease categories, enabling precise detection and segmentation without manual preprocessing. Advanced architectures like U-Net and SegNet have also shown promising results in reconstructing infected areas, with U-Net achieving an average precision of 97.5% and recall of 96.2% for identifying diseased regions.

Other CNN variants, such as EfficientNetV2B2 [4], explored by Anjan Debnath, used a dataset of 10,000 tomato leaf images to achieve a test accuracy of 98.96% and an F1-score of 98.5%. DenseNet121 [8], developed by Naresh K. Trivedi, and a hybrid CNN model combining VGG and Inception modules by Maria Vasiliki Sanida [6], both trained on the PlantVillage dataset of 18,160 images, demonstrated high effectiveness in classifying tomato leaf diseases with validation accuracies of 99.17% and F1-scores of 99.1%.

Thippa Reddy G's PCA-WOA-DNN model [3] combined dimensionality reduction (PCA) with the Whale Optimization Algorithm (WOA) for feature selection before classification using a deep neural network (DNN). This approach, trained on the PlantVillage dataset, significantly improved classification performance, achieving 94% testing accuracy and an F1-score of 93.8%. The combination of PCA and WOA enhanced feature selection and reduced overfitting, resulting in more reliable outcomes in real-world scenarios.

Ensemble models using EMA (Exponential Moving Average) and EWGO (Enhanced Weighted Gradient Optimizer), proposed by V. Pandiyaraju [7], were also trained on the PlantVillage dataset and achieved an average accuracy of 98.7%, a recall of 98.6%, and an F1-score of 98.4%. Mohammed Saeed Alzahrani's Vision Transformer (ViT) model [9] combined transformer capabilities with computer vision, achieving an accuracy of 98% and an F1-score of 97.8%, highlighting its effectiveness in disease detection and classification.

Bhandari et al. [5] developed the BotanicX-AI model using the EfficientNetB5 architecture, trained on a dataset of 11,000 images from Kaggle. By incorporating explainability techniques like GradCAM and LIME, the model achieved an F1-score of 98.9%, providing high classification accuracy along with enhanced interpretability, making it suitable for real-world agricultural applications.

Irmak and Saygılı [2] proposed a CNN-based approach for tomato leaf disease classification, using a dataset of 18,345 images from Kaggle, organized into ten classes. The CNN model outperformed classical machine learning models in all scenarios, achieving 99.5% accuracy in binary classification, 98.50% in 6-class classification, and 97% in 10-class classification, demonstrating its superior performance in handling multi-class problems.

These advancements in segmentation and classification represent significant progress in managing agricultural diseases, particularly in complex environments, thereby improving the efficiency and effectiveness of smart agriculture systems.

3 Comparative Analysis of Models for the Classification and Segmentation of Tomato Diseases

The Table 1 provides an overview of various studies on the detection of tomato leaf disease indicators. The diseases considered include Bacterial Spot, Early Blight, Late Blight, Leaf Mold, Septoria Leaf Spot, Mosaic Virus, and Yellow Curl Leaf. Each checkmark signifies that a particular study has addressed the detection of a specific disease. The table offers a comprehensive view of research coverage for major tomato leaf diseases, helping to identify which diseases have been examined in each reference.

Table 1. Overview of Research on Detection of Tomato Leaf Disease Indicators

Disease indicators							
References	Bacterial Spot	Early Blight	Late Blight	Leaf Mold	Septoria Leaf Spot	Mosaic Virus	Yellow curl leaf
[1]					√	√	√
[2]	√	√	√	√	√	√	
[3]	√	√	√	√	√	√	
[4]	√	√	√	√	√	√	
[5]	√	√	√	√	√	√	
[6]	√	√	√	√	√	√	
[7]	√	√	√	√	√	√	
[8]			√				√
[9]	√	√	√	√	√	√	

The Table 2 provides an overview of various solutions and results from studies focused on detecting tomato leaf diseases. The table outlines the methods employed in each study, such as Convolutional Neural Networks (CNN), Principal Component Analysis (PCA), Whale Optimization Algorithm (WOA), EfficientNet, and hybrid models. It also details the performance metrics of these methods, including classification accuracy, loss rates, and precision and all performance metrics in detecting diseases like Bacterial Spot, Early Blight, Late Blight, Leaf Mold, Septoria Leaf Spot, and Mosaic Virus.

The Table 3 provides a summary of the proposed solutions and results for the detection of tomato leaf diseases, specifically focusing on Mosaic Virus and Yellow Curl Leaf. The adaptive K-means algorithm, using the Davies-Bouldin (DB) index to determine the optimal number of clusters, showed improved segmentation capabilities. The results demonstrated an F1 score of 0.982, which significantly outperformed traditional methods like K-means (0.894), DBSCAN (0.867), and Mean Shift (0.805). Additionally, the adaptive K-means achieved a lower entropy value of 0.118 compared to K-means (0.317), DBSCAN (0.396), and Mean Shift (0.485), indicating better segmentation quality and less uncertainty. The method also provided higher accuracy in complex backgrounds and better separation of tomato leaves compared to traditional clustering approaches, highlighting its effectiveness in handling challenging segmentation scenarios.

The Table 4 provides a summary of the proposed solutions and results for the detection of Late Blight and Yellow Curl Leaf diseases in tomato leaves. The approach utilized infrared, red, green, and blue (IRGB) color images of tomato leaves as input. Using a CNN-based segmentation technique, a segmented mask was generated, followed by dividing the image into smaller regions for detailed classification. The proposed CNN model achieved an accuracy of 98.49%, with precision, recall, and F1-score values of 99%, 99%, and 98.99% respectively. Additionally, the proposed CNN model outperformed other well-known models, such as MobileNet (66.75%), VGG16 (79.52%), and InceptionV3 (64.25%), demonstrating its superior effectiveness in identifying tomato leaf diseases.

Table 2. Comparative Analysis of Deep Learning Models for Tomato Leaf Disease Detection

Disease indicators							
References	Bacterial Spot	Early Blight	Late Blight	Leaf Mold	Septoria Leaf Spot	Mosaic Virus	Yellow curl leaf
	Proposed Solutions and Results						
[2]	• Binary Classification: CNN accuracy 99.5%, SVM 92.5%, kNN sensitivity 98% • 6-Class Classification: CNN accuracy 98.5%, SVM accuracy 83%, sensitivity 94%, specificity 98.6% • 10-Class Classification: CNN accuracy 97%, SVM accuracy 81.4%, specificity 98.9%, sensitivity 92.7% • Comparison: CNN outperformed SVM, kNN, ELM across all tasks						
[3]	• PCA Dimensionality Reduction: Training accuracy 99%, testing accuracy 90%, reduced training time • WOA Optimization: Improved testing accuracy to 94% post-PCA • DNN without Reduction: Training accuracy 99%, testing accuracy 86%, 5% loss rate (overfitting) • Hybrid PCA-WOA-DNN: Best results—training accuracy 99%, testing accuracy 94%, loss rate 1.8%, reduced training time • Summary: Highest accuracy 94%, minimized loss rate 1.8%, improved efficiency						
[4]	• EfficientNetV2B2 Performance: Avg training accuracy 99.02%, validation 99.22%, test 98.96%; best fold test accuracy 99.5% • Split Method Results: Training accuracy 99.93%, validation 100%, test 100% • Performance Metrics: Precision, recall, F1-score all 100% • Comparison with Other Models: Outperformed InceptionV3 (+8.73%), EfficientNetB3 (+0.73%), SE-ResNet50 (+2.99%).						
[5]	• EfficientNetB5 Performance: Training accuracy 99.84%, validation 98.28%, test 99.07% • Explainability: GradCAM and LIME for model transparency • Model Comparison: EfficientNetB5 test accuracy 99.07%, outperforming MobileNet (94%), Xception (95.32%), VGG16 (93.35%), ResNet50 (96.03%), DenseNet121 (96.3%) • Metrics: Precision, recall, F1-score near 100%, AUC-ROC score 1.0						

(*continued*)

Table 4. Performance Comparison of Models for Late Blight and Yellow Curl Leaf Detection in Tomato Leaves

Disease indicators		
References	Late Blight	Yellow curl leaf
	Proposed Solutions and Results	

(*continued*)

Table 4. *(continued)*

Disease indicators		
References	Late Blight	Yellow curl leaf
[8]	• Input of Infrared, Red, Green, and Blue (IRGB) color images of tomato leaves. • Generation of the segmented mask Mveq using CNN-based segmentation. • Division of the image into smaller regions for classification. • Proposed CNN: Accuracy 98.49%, precision 99%, recall 99%, F1-score 98.99% • Comparison: Outperformed MobileNet (66.75%), VGG16 (79.52%), InceptionV3 (64.25%)	

4 Results

The performance of various deep learning models for tomato leaf disease classification was evaluated using multiple performance metrics, including accuracy, precision, recall, F1-score, and loss rate. The results demonstrate significant improvements in the efficiency of classifying images of tomato leaves affected by different diseases, utilizing advanced image processing techniques and Convolutional Neural Network (CNN) architectures. Some classifications are illustrated as follow:

- **Classification of Diseases Using IRGB and CNN-Based Segmentation**

The use of IRGB images combined with CNN-based segmentation achieved a classification accuracy of 98.49%, precision of 99%, recall of 99%, and an F1-score of 98.99% for detecting Late Blight symptoms. This model outperformed other architectures such as MobileNet (accuracy: 66.75%), VGG16 (accuracy: 79.52%), and InceptionV3 (accuracy: 64.25%) [8].

- **Image Segmentation Using Adaptive K-Means**

Another model, based on adaptive K-means clustering, was applied to segment tomato leaves affected by Mosaic Virus. This model, enhanced with validity indices such as the Davies-Bouldin Index and the Calinski-Harabasz Index, demonstrated robust segmentation of tomato leaf images, surpassing traditional methods like DBSCAN, Mean Shift, and ExG-ExR [1]. The metrics showed an F1-score of 0.982 compared to lower values achieved by traditional methods (K-means: 0.894, DBSCAN: 0.867, Mean Shift: 0.805). The entropy value for adaptive K-means was 0.118, indicating better segmentation quality compared to K-means (0.317), DBSCAN (0.396), and Mean Shift (0.485).

Table 2. (*continued*)

Disease indicators							
References	Bacterial Spot	Early Blight	Late Blight	Leaf Mold	Septoria Leaf Spot	Mosaic Virus	Yellow curl leaf
[6]	• Hybrid CNN (VGG + Inception): Training accuracy 99.83%, validation 99.17%, test 99.17% • Comparison: Outperformed InceptionResNet (98.4%) and ResNet152 (97.3%) • Metrics: Precision 99.13%, recall 99.23%, F1-score 99.17%, AUC 99.56% • Class-wise Performance: 100% precision, recall, F1-score for some classes						
[7]	• Adaptive Ensemble (VGG16 + NASNet): Accuracy 98.7%, error rate 4% • Metrics: Precision 97.9%, recall 98.6%, F1-score 98.7%, AUC 99.97% • Comparison: Outperformed traditional classifiers in accuracy, precision, and F1-score						
[9]	• DenseNet121: Training accuracy 99.88%, test 99%, precision 99%, recall 99%, F1-score 98.99% • ResNet50V2: Training accuracy 99.49%, test 95.6%, precision 95.6%, recall 95.6%, F1-score 95.59% • ViT: Training and test accuracy 98%, precision 98%, recall 98%, F1-score 98%						

Table 3. Image Segmentation Techniques for Tomato Leaf Disease Detection

Disease indicators		
References	Mosaic virus	Yellow curl leaf
	Proposed Solutions and Results	
[1]	• Adaptive K-means: Improved segmentation, DB index for cluster number • Metrics: F1 score 0.982 (vs. K-means 0.894, DBSCAN 0.867, Mean Shift 0.805) • Entropy: Lower entropy 0.118 (vs. K-means 0.317, DBSCAN 0.396, Mean Shift 0.485) • Comparison: Higher accuracy in complex backgrounds, better leaf separation than traditional methods	

- **Hybrid CNN Models and Deep Networks**

The integration of hybrid CNN models using VGG blocks and Inception modules achieved a training accuracy of 99.83%, validation accuracy of 99.17%, and an F1-score of 98.7%. The use of an Exponential Moving Average (EMA) function further enhanced performance, achieving a final accuracy of 98.7% with a loss rate below 4% and a precision of 97.9% [6][7].

- **DenseNet and Vision Transformer (ViT) Models**

The DenseNet121 model achieved a training accuracy of 99.88%, testing accuracy of 99%, precision of 99%, recall of 99%, and an F1-score of 98.99%. The ResNet50V2 model achieved 99.49% training accuracy, 95.6% testing accuracy, precision of 95.6%, recall of 95.6%, and an F1-score of 95.59%. The Vision Transformer (ViT) model demonstrated solid performance, with both training and testing accuracy of 98%, precision of 98%, recall of 98%, and an F1-score of 98% [9].

- **Classification Using PCA and WOA**

The integration of Principal Component Analysis (PCA) and the Whale Optimization Algorithm (WOA) achieved a training accuracy of 99%, testing accuracy of 94%, precision of 94%, recall of 94%, and an F1-score of 94%, demonstrating the effectiveness of these methods in classifying tomato leaf diseases without compromising accuracy [2][3].

These results demonstrate not only qualitative improvements in segmentation and classification but also significant quantitative advancements across all metrics, reinforcing the models' effectiveness in real-world scenarios for detecting tomato leaf diseases.

5 Discussion

The analysis of various deep learning models for tomato leaf disease detection reveals significant variations in performance, particularly regarding classification accuracy. While some models demonstrate exceptionally high accuracy, real-world deployments introduce technical challenges that can impact model reliability and effectiveness.

One of the primary challenges faced by these models is the variation in environmental conditions. Changes in lighting conditions, such as fluctuating sunlight or shadows, can significantly affect image quality, thereby impacting model accuracy. Models like DenseNet121, which achieved 99.88% training accuracy and 99% testing accuracy, may struggle to maintain this level of performance under varying lighting conditions, particularly in open-field environments where lighting is not controlled. This limitation highlights the need for robust preprocessing techniques and adaptable models that can compensate for such variations.

Another technical challenge involves the angles at which images are captured. Different camera angles can introduce variability in the appearance of leaf features, which can affect classification accuracy. The CNN model combined with PCA and the Whale Optimization Algorithm (WOA), which achieved 99.5% accuracy, might face difficulties in generalizing to images captured from inconsistent angles. The use of image augmentation during training could help improve the model's ability to handle these variations, thereby enhancing its robustness in practical applications.

Image quality also plays a crucial role in the effectiveness of disease detection models. Factors such as resolution, noise, and occlusions can degrade image quality, making it difficult for models to accurately segment and classify diseased areas. Models like ResNet50V2, which showed a decline in testing accuracy (95.6%) compared to training accuracy (99.49%), may be particularly sensitive to poor image quality. Techniques such as noise reduction, resolution enhancement, and advanced filtering could be beneficial in mitigating these effects.

The Hybrid CNN model using VGG and Inception modules, which achieved 99.83% training accuracy and 99.17% validation accuracy, also highlights the importance of feature extraction across multiple scales. However, in real-world scenarios, leaf occlusion by other plant parts or environmental elements may hinder the model's ability to extract relevant features effectively. Addressing these issues through improved image preprocessing and the use of attention mechanisms could enhance model performance under challenging conditions.

In conclusion, while models like DenseNet121 and the CNN model with PCA and WOA demonstrate high performance in controlled conditions, their deployment in real-world agricultural environments presents unique challenges. Addressing limitations related to lighting variability, image capture angles, and quality is crucial for ensuring the reliability and accuracy of these models in practical settings. Future work should focus on developing adaptive techniques and robust preprocessing strategies to mitigate these challenges, ultimately improving the effectiveness of tomato leaf disease detection in precision agriculture.

6 Conclusion et Perspectives

In this study, we focused on the classification of tomato leaf diseases using deep learning models, achieving promising results with models such as DenseNet121 and the CNN model combined with PCA and the Whale Optimization Algorithm (WOA), which achieved the highest classification accuracies of 99.88% and 99.5%, respectively. These models demonstrated significant potential for real-world applications in precision agriculture, where early detection and classification of plant diseases are crucial for preventing crop losses.

Although our primary interest was in classification, we also explored image segmentation techniques, particularly in the detection of diseases like Mosaic Virus, where segmentation algorithms such as adaptive K-means clustering were applied to improve the accuracy of disease detection. The exploration of segmentation methods opens new avenues for future research, as accurate segmentation can enhance feature extraction and further improve classification outcomes.

Looking ahead, future research should address some of the technical challenges encountered in real-world agricultural settings. These include variability in lighting conditions, differing image capture angles, and inconsistencies in image quality. Solving these issues will be essential to ensuring the robustness and reliability of disease detection models in diverse environments.

In addition, we aim to combine the strengths of the two top-performing models, DenseNet121 and the CNN model with PCA and WOA, to develop a hybrid model that can deliver even higher precision in disease detection. By merging the dense connectivity and feature propagation of DenseNet121 with the optimized feature selection capabilities of the PCA-WOA model, we expect to improve the robustness and generalizability of the system across a wider range of disease conditions and datasets.

This future work will not only enhance the classification of tomato leaf diseases but also integrate advanced segmentation techniques while addressing the aforementioned technical challenges, creating a more comprehensive disease detection framework. Such a hybrid approach could revolutionize disease monitoring in agriculture, providing more reliable and efficient tools for early diagnosis and intervention.

References

1. Tian, K., Li, J., Zeng, J., Evans, A., Zhang, L.: Segmentation of tomato leaf images based on adaptive clustering number of K-means algorithm. Comp. Electron. Agr. **165**, 104962 (2019). https://doi.org/10.1016/j.compag.2019.104962
2. Irmak, G., Saygılı, A.: A novel approach for tomato leaf disease classification with deep convolutional neural networks. J. Agr. Sci. (Tarim Bilimleri Dergisi) **30**(2), 367–385 (2024). https://doi.org/10.15832/ankutbd.1332675
3. Gadekallu, T.R., et al.: A novel PCA–whale optimization-based deep neural network model for classification of tomato plant diseases using GPU. J. Real-Time Image Proc. **18**, 1383–1396 (2021). https://doi.org/10.1007/s11554-020-00987-8
4. Debnath, A., et al.: A Smartphone-Based Detection System for Tomato Leaf Disease Using EfficientNetV2B2 and Its Explainability with Artificial Intelligence (AI). Sensors **23**(21), 8685 (2023). https://doi.org/10.3390/s23218685
5. Bhandari, M., Shahi, T.B., Neupane, A., Walsh, K.B. : BotanicX-AI: Identification of Tomato Leaf Diseases Using an Explanation-Driven Deep-Learning Model. Journal of Imaging **9**(2), 53 (2023). https://doi.org/10.3390/jimaging9020053
6. Sanida, M.V., Sanida, T., Sideris, A., Dasygenis, M.: An Efficient Hybrid CNN Classification Model for Tomato Crop Disease. Technologies **11**(1), 10 (2023). https://doi.org/10.3390/technologies11010010
7. Pandiyaraju, V., et al.: Improved tomato leaf disease classification through adaptive ensemble models with exponential moving average fusion and enhanced weighted gradient optimization. Frontiers in Plant Science **15**, 1382416 (2024). https://doi.org/10.3389/fpls.2024.1382416
8. Trivedi, N.K., et al.: Early detection and classification of tomato leaf disease using high-performance deep neural network. Sensors **21**(23), 7987 (2021). https://doi.org/10.3390/s21237987
9. Alzahrani, M.S., Alsaade, F.W.: Transform and deep learning algorithms for the early detection and recognition of tomato leaf disease. Agronomy **13**(5), 1184 (2023). https://doi.org/10.3390/agronomy13051184

A Review on Advancement in PEM Fuel Cell Diagnosis Based on Machine Learning Techniques

Hichem Kahia[1(✉)], Selma Boumerdassi[2], Aissa Belmeguenai[5], Abderrahmane Herbadji[3], and Djamel Herbadji[4,5]

[1] Laboratoire de Modelisation des Systemes Energetiques LMSE Univ of Biskra, BP145, 7000 Biskra, Algeria
hicham.kahia@univ-biskra.dz

[2] Conservatoire National des Artes et Metiers, 292 Rue Martin, 75141, 90400 Paris Cedex 03, France
selma.boumerdassi@cnam.fr

[3] Laboratoire de Recherche en Electronique de Skikda, Université 20 Août Skikda, BP 26 Route d'El-hadaeik, 21000 Skikda, Algeria

[4] Department of Electronics, University of Setif-1, 19000 Setif, Algeria

[5] Laboratory of Signals and Systems, Institute of Electrical Engineering and Electronics, University M'Hamed Bougara of Boumerdes, Boumerdes, Algeria

Abstract. Proton exchange membrane (PEM) fuel cell has attracted much attention due to its high efficiency and environmental friendliness, whose only reaction product is water. The main obstacles to large-scale deployment are premature fuel cell failures that limit the PEM fuel cell lifetime, which brings us back to many difficulties. In this study, we have defined and discarded the main methodological approaches to identify and diagnose FC. In this context, we have discussed the possible existing PEM fuel cell diagnosis methods that provide insight into FC performance, with the aim of improving diagnostic techniques and to deepen the understanding of the diagnostic behavior of PEMFC.

Keywords: PEM Fuel Cells · (SOH) · AI technology · Neural network model · (EIS)

1 Introduction

It's well known that fuel cell produces water when it delivers a current. The produced water is generally too much and therefore it will slow the diffusion of gases and distribute them well over the diffusion layers. Here, water management issue is particularly important in PEMFC Park and Caton (2008); Ge et al. (2006).

F. Hacène et al. (Eds.): MLN 2024, LNCS 15540, pp. 180–194, 2026.
https://doi.org/10.1007/978-3-032-00552-6_13

The state and transport of water always have a significant effect on the output characteristics and reliability of the fuel cell under various conditions (e.g. drying or flooding start and normal operation) in each of the single cells, water distribution in the membrane as well as the difference in water content in the anodic and cathodic compartments, observing the onset and slow propagation of the FC dehydration, and studying the effect of water supply in reactive gases on the electrical performance from the local current density measurement and the measurements of the fields. PEM fuel cell diagnostics has been widely adopted in the literature, mainly due to the simplicity of identifying the water condition in the fuel cell. The diagnosis of PEM fuel cell failure modes (flooding and drying) is very important for the normal operation of PEMFC, because dehydration hinders the supply of protons into the membrane, and therefore the performance will be limited. Additionally, water accumulation usually leads to flooding inside the fuel cell, this flooding can prevent the transport of reactants and result in gas starvation. The existing PEM fuel cell diagnosis methods such as analytical, semi-empirical, empirical, physical techniques are remarkably affected by several factors. For instance, computationally expensive, the need for experimental and invasive instrumentation, large number of characteristic quantities, slowness of the temporal resolution.

To this end, in this work, we propose to deploy artificial intelligence for water management failure in PEMFCs which directly affects the sustainability and stability of fuel cells. The main idea behind the use of artificial intelligence for PEMFC diagnosis is that it exhibits high sensitivity in identifying specific parameters and is capable of predicting the voltage response under a sudden change in relative humidity.

2 Overview of a PEFC Single Cell (The Heart of the Fuel Cell)

Polymer exchange membrane fuel cell (PEMFC) represents an alternative energy conversion for stationary and automotive applications. The thickness of the electrolyte mombran is about ten micrometers (from 20 to 200 ?m as a general rule). A low water content decreases the ionic conductivity and increases its gas permeability. On the other hand, too high water contents can cause mechanical stresses due to the swelling of the membrane O'Hayre et al. (2006); Baschuk and Li (2001).

The membrane is placed between two electrodes which constitute the catalyst layer (CL), seat of the oxidation-reduction half-reactions. The two active layers associated with the membrane constitute the Membrane Electrodes Assembly (MEA) Qi (2013) Fig. 1.

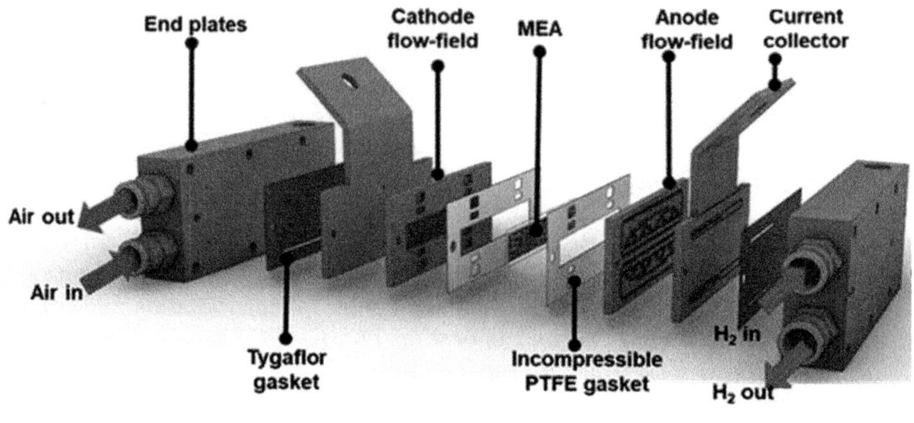

Fig. 1. Fuel cell composition, aluminum end plates, current collectors and flow fields

The AME, which is the heart of the PEMFC, is sandwiched between two gas diffusion layers (GDL for Gas Diffusion Layer) Spiegel (2007). The diffusion layer provides several functions: (i) serves to supply reactants (H2 and O2/air to the anode and cathode respectively), (ii) provides electrical conduction between the bipolar plates and the active areas, (iii) evacuates the heat produced in the electrodes (resulting from reactions and ohmic losses), (iv) evacuates the water produced at the cathode towards the channel far from the electrodes, in order to prevent their flooding and drying (v) provide mechanical support for the MEA Sharaf and Orhan (2014); Korner et al. (2015); El-Kharouf et al. (2012).

The external fluidic circuits are connected to the cell at the terminal plates and the reactant gases are distributed to the MEA on the surface of the bipolar plates via channel setched on them. Moreover, the low weight and high performance are important for manufacturing high-volume plats to be practical in industrial applications.

To design the fuel cell corresponding to the book of terms of a given project, the designer has two degrees of freedom: The number N of the fuel cells placed in series and the surface A of a cell. The overall system (core cell and auxiliaries) is called the fuel cell system. The auxiliaries of a fuel cell consume a significant part of the energy produced by it, thus causing the degradation of the efficiency of the system as shown in following Fig. 2. Dubas (2006).

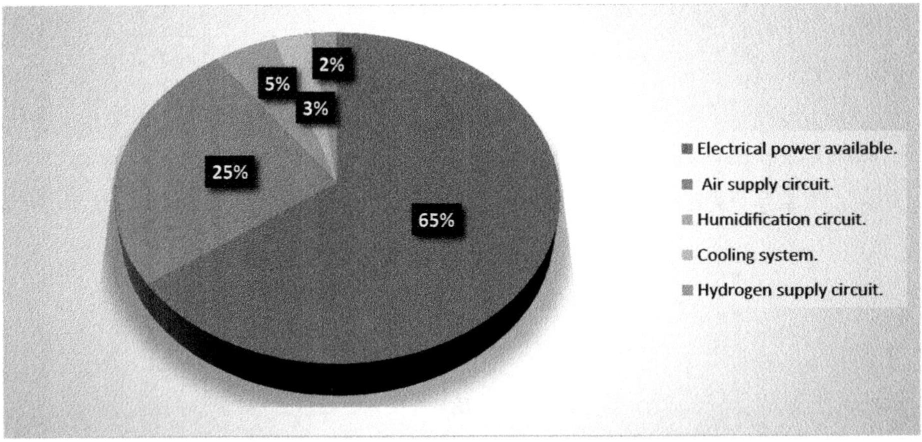

Fig. 2. Division of electricity consumption in a fuel cell system

The role of the humidification circuit is to humidify the gases entering the cell, generally from the water produced by the cell, recovered by means of a condenser. Drying or over-wetting the membrane decreases electrical energy production and limits battery life.

3 Degradation and Correlation with the Current Local Distribution

The actual cell potential is decreased from the ideal potential due to several types of irreversible losses Zhang et al. (2013), as shown in Fig. 3 Hou et al. (2019). Because of these different irreversibilities, the polarization curve of a fuel cell (cell voltage as a function of current density) exhibits a non-linear electrochemical behavior Zhang et al. (2006, 2014); Cheng et al. (2007). This characteristic is called the law of polarization. "Dynamic" reading, by current sweeping of high amplitude, at low frequencies, makes it possible to obtain very good results, with more information than point-to-point reading. Here, time, current and voltage information are thus retrieved, in much greater values than the traditional reading. Compared to the point-to-point survey, this method allows a fairly easy reproducibility of the tests and is much less dependent on the user. It also allows to obtain a very large number of points. The notable difference between the point-to-point reading and the dynamic plotting is the time spent at each current value. In the case of the point-to-point survey, each point was surveyed by varying the current, then waiting for a stabilization time. In the case of dynamic tracing, the current only evolves on principle. As shown in Fig. 3, the pressure was automatically regulated to 2 bar and the H_2 and O_2 flows were set to satisfy the max current. The pressure and flow conditions were the same for the manual reading or current sweep reading tests Fontes (2005).

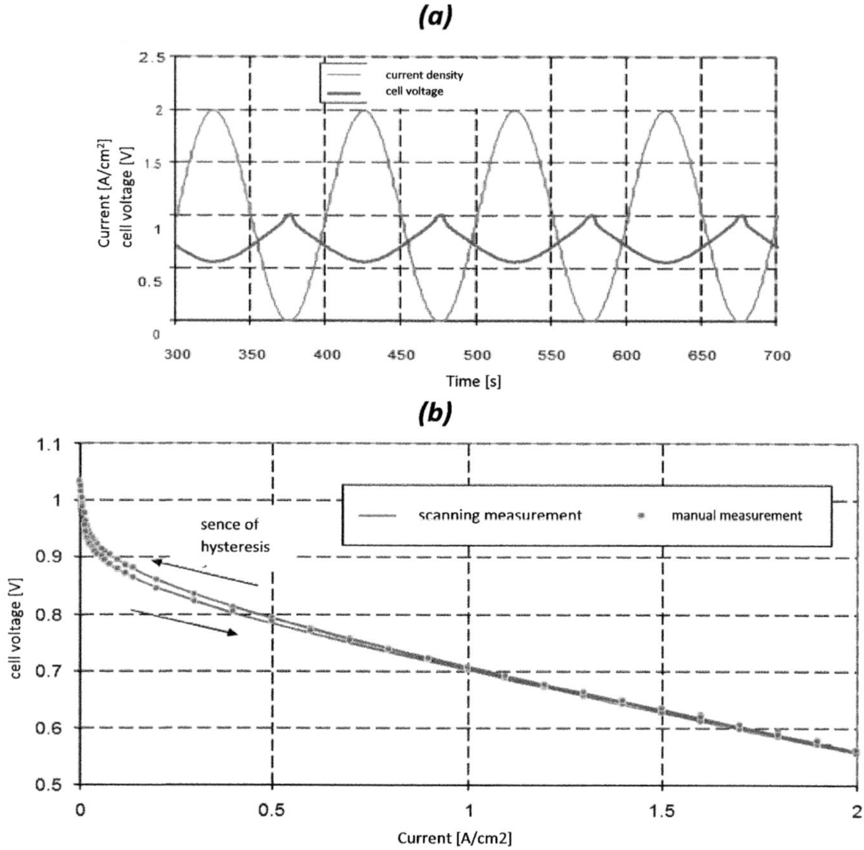

Fig. 3. The sinusoidal current sweep at 10 mHz and the Comparison of the V I curves of a point-to point reading "by hand" and that obtained by Monocell sweep supplied with H2/O2

The degradation mechanisms indirectly impact the current distribution in the cell in different ways; either by the local modification of the resistance of the fuel cells, or by reducing the active surface of the cell due to the degradation of the catalyst. In the case of hydrogen depletion, the current density distribution of a cell subjected to different stoichiometric measurements of hydrogen (from 0.2 to 1) has been studied Liang et al. (2011), and the current measurement is performed when using a segmented cathode (segment 1: gas inlets and segment 5: gas outlet) as shown in Fig. 4.

In order to achieve optimal efficiency, it is necessary to understand, detect, diagnose and isolate each of these failure modes in the PEMFC: (i) Carbon corrosion often occurs at the cathode depending on the reaction, this reaction mainly occurs at high potentials. It leads to a loss of carbon mass. Endurance tests have shown a mass loss of up to 15% under 0.95 V (open circuit voltage) and

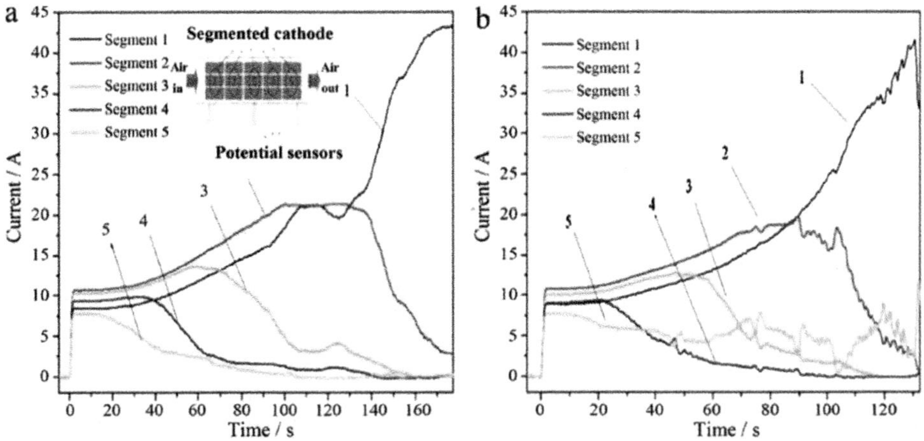

Fig. 4. Evolutions of local current densities for a segmented cell: (a) H2 stoichiometry at 1,(b) H2 stoichiometry at 0.9

up to 5% under 0.75 V. The carbon is then consumed on the anode side Kahia et al. (2020). This phenomenon is limited in advance and is favored at the cathode. (ii) Dissolution and migration of the platinum catalyst, this translates concretely into an irreversible loss of active surface and therefore of power deliverable by the fuel cell Yasuda et al. (2006); Wang et al. (2006). To overcome the problems related to platinum, in the works of Zhang et al. (2007) Authors suggested to replace platinum with platinum/gold clusters. (iii) Membrane degradation, It leads to the appearance of free radicals (RH), that reduce the amount of beneficial substances. A slow degradation of the membrane, for example, reduces its ability to transport ions, resulting in an increase in the electrical resistance of the fuel cell (therefore an increase in losses). In the case of storage, or not used for several weeks or several months, the membrane risks drying out too much.

4 Modeling and Characterisation Methodologies

There is a significant amount of methodological approaches that help to identify and diagnose a FC system. These approches can be classified into five families: knowledge models, semiempirical models, black box models, empirical approaches and finally information processing techniques. The above-mentioned methodologies can be classified into two types: (static methods and dynamic methods). The static approach is particularly interesting for making a technological choice concerning, for example, the catalyst or for sizing fuel cell components. While dynamic methods are preferable when one wishes to analyse transient phenomena associated to a sudden change in a setpoint or a parameter.

4.1 Mechanistic Model

The first family of model is known under different names such as: knowledge model, mathematical model, physical model, or analytical model. This category of model calls upon precise relations describing in a physical way the various interactions between the input variables and the output vector according to the surrounding conditions (materials, boundary conditions, initial conditions). Even if more or less strong hypotheses make it possible to simplify the approach, the disadvantages of analytical models lie in the slowness of the temporal resolution of the equations, especially when the model is dynamic and in 3 dimensions (study on the 3 axes), as well as than in the validation of the model due to the large number of characteristic quantities to be estimated or measured. Finally, the need for experimental and sometimes invasive instrumentation is important in order to determine the characteristic quantities. The use of equivalent electrical models involves passive electrical components to model the dynamic behavior of the fuel cell. This methodology is based on an experimental and multiparametric study allowing to determine the values of the model elements. Although mainly coupled with electrochemical impedance spectroscopy (EIS), some works has been undertaken using the current interruption (IC) method Rubio et al. (2007). The advantage of this approach is the physical connection to the passive components of the model at a relatively low computational cost (parameter determination). Therefore, parameters cannot be determined in a single way as the impedance spectrum or current voltage reading that can be modeled by different equivalent circuits. However, using EIS as an equivalent model and complex nonlinear least squares algorithm allowed detection and identification of 3 defects (CO poisoning, immersion of electrodes and membrane drying) by varying model parameters Latham (2004). On the contrary, there is a method to use the algorithm of multiparametric resolution, the use of the method by IC to wedge its model of Randles. The advantage of his approach lies in the fact that the solution found is unique with respect to the plot. Its disadvantage is that the IC method requires a fast acquisition system.

4.2 Black Box Model

The final interest of the black box approach is clearly to minimize the computation time linked to the model in order to be able to perform a control-command action on the system. The main disadvantage of this type of modeling is that there is no physical link between the model and the reality of things. There are studies that use fuzzy logic and residue analysis as diagnostic vectors for two faults Hissel et al. (2004): membrane drying out and water or nitrogen accumulation in the anode compartment. The residual values are calculated by considering the difference between the experimental operating point (current-voltage) and the current-voltage couple resulting from a reference polarization curve. The open-loop model is used to reduce computation time, which is combined with a genetic algorithm to increase detection sensitivity. The decision step is entrusted to an empirical probability threshold coupled with a logic circuit. This approach is interesting from a practical point of view since it only requires two low-cost sensors (current and voltage). While, authors in Fennie et al. (2001) have developed a model based on fuzzy logic with 3 inputs and one output according to the Standard additive model. In addition to the polarization curve, it uses electrochemical impedance spectroscopy to specifically identify the drying defect. In this perspective, the authors use the increase in the impedance amplitude as well as a decrease in the performance of the polarization curve to distinguish between drying and flooding respectively (fuzzy classification). The model is trained by an algorithm that classifies the health status of the generator into several levels (at least 3, one for each error). Although this method offers a good level of fault detection, it requires a large database to get an interesting level of detection. Another approach in Nitsche et al. (2004) uses neural networks to diagnose any real-time change in the polarization curve due to aging effects, impurities contained in reactive gases and static or dynamic operation. The diagnosis is performed by comparing the polarization curve estimated by a model (through static or dynamic measurements) with a reference polarization curve. The polarization curve is estimated by an evolutionary learning algorithm (permanent learning) which transcribes the measured operating points into polarization curves. Neural networks have also been employed to detect the presence of contaminants in feed gases Saber et al. (2019).

4.3 Semi-empirical Model

Semi-empirical models occupy an intermediate place between physical models and behavioral models known as black boxes. This type has been used to diagnose the state of the fuel cell through the so-called current interruption method. Researchers have suggested numerous models to diagnose the PEMFC. For instance, the model developed by Forrai et al. (2005) is a simplification (reduction of the order) of a physical model using equivalent electrical circuits. The same method was used by Nakajima et al. (2008) for diagnosing the presence of CO. Another semi-empirical model has been developed by Li et al. (2008) to study the effects of toluene contamination on the polarization curve. Finally, authors in Sugiura et al. (2006) used a semi-empirical approach combined with the current interruption method to detect the presence of the flooding phenomenon. The developed equivalent circuit is made up of a transmission line of three RC cells allowing modeling of the nonlinear equation linking the voltage to the current, the flaw detection is observed by varying certain parameters in the model.

The semi-empirical model has many advantages (e.g. increased convergence speed compared to physical models, preservation of part of the physical relations, etc.), and its interest in diagnostic context is informative and interesting.

4.4 Signal and Information Processing Approaches

4.4.1 Signal Approaches

The development of a physical, semi-empirical or black box model for diagnosis is relatively tedious, non-intuitive, costly in financial terms and in development time. To overcome these difficulties, the "signal" approach has been proposed as an alternative to previously mentioned approches. In this case, it is actually a matter of finding one or more specific measurements (frequency, voltage, pressure, etc.) to characterize one or more defects it worth noting that. This approach has been widely implemented for fuel cell diagnostics.

An original method was proposed for detecting a possible accumulation of excessive water in the anode and cathode compartments Mérida Donis (2002). The authors use the frequency response of the pressure sensor followed by a rung of current applied to the FC to estimate the water content in the cell (if there is for example flooding) using a local current density measurement in a segmented fuel cell. Furthermore, electrochemical impedance spectroscopy (EIS) is a technique capable of characterizing several physical and chemical processes provided that their time constants are distinct. The effect of drying being visible on the impedance spectrum in the frequency band 100 kHz–0.5 Hz. The effect of flooding being visible on a low frequency range starting at 100 Hz to 0.5 Hz. The identification of frequency zones allowed the diagnosis and the determination of the defect. However, the measurement of several low-frequency points remains problematic. This problem was avoided by using three specific frequencies: 50 Hz, 500 Hz, and 5000 Hz to detect and identify poisoning, drying out and flooding faults Mérida Donis (2002); Mérida et al. (2006).

4.4.2 Information Processing

The information processing approach is implemented in a way (logical, statistical or even probabilistic) based on the voltage distribution of the unit cells of the generator using a stochastic approach. The methodology allowed for a clear distinction between the different operating modes. Moreover, this approach has the advantage of remaining simple from the algorithmic point of view and thus, one can easily implement this method on embedded systems to get a real-time diagnosis Hernandez (2006); Hernandez et al. (2006).

The probabilistic approaches have also been considered for the diagnosis of FC Hernandez (2006). This graphical method identifies or finds the logical cause and effect relationships present in the database through the various variables defined in the Bayesian network with five input parameters (voltage, current, electrical power, temperature and hydrogen pressure at the input of the stack) enabling the diagnosis of many faults. In addition, the graphical model of the Bayesian network was constructed using discriminating variables, observations, and diagnostics, these variables can only be defined by the network manufacturer.

Figure 5 summary of the main characterization and various experimental methods for performing diagnostics on the PEM fuel cell.

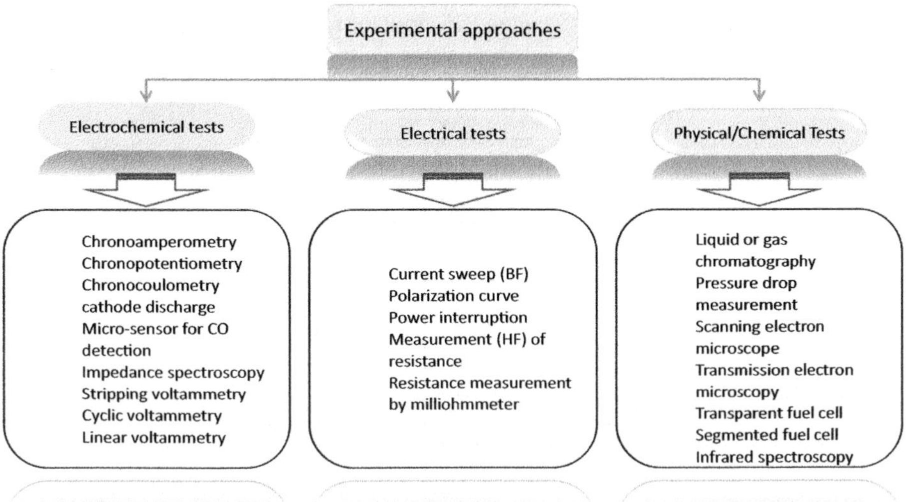

Fig. 5. Summary of the main characterization and diagnosis methods applicable to a FC system or one of its elements

5 Learning Rules and Machin Learning Based PEMFC Diagnosis

A simple way to avoid overfitting consists in regularization or introducing a penalty, in the scale to be optimized. when the value of the parameter is increased, the inputs weights of the neurons less and can take on chaotic values, consequently, helping to limit the risks of over-learning as illustrated in the following diagram Fig. 6.

Recently, artificial neural networks (ANNs) based diagnosing method is widely employed. Many researchers have proposed the use of neural network method to diagnosis the PEM fuel cells Steiner et al. (2011); Jemeı et al. (2003); Cadet et al. (2014); Kim et al. (2012); Kim et al. (2012). For instance:

The PEMFC approach of controlling a dynamic neural network was investigated by the authors of Hatti and Tioursi (2009). Their research's goal is to track the changes made to the system, commencing with the addition of a delay line to the (NNT)'s input to create a dynamic (NNT) control model.

For the purpose of categorizing the health states (SOH) of the fuel cells (flood, normal functioning, and drying out), Steiner et al. Steiner et al. (2011) suggested a model employing (NNT) and they showed the distinction between the normal functioning designed by a neural network and the real functioning.

In their investigation of diagnostic methods, Jonghoon et al. Kim et al. (2012) used pattern recognition model identification (FCOV) based on Hamming NNT application. to choose the cell's (ΔRd) value and use it to determine the condition of the cell.

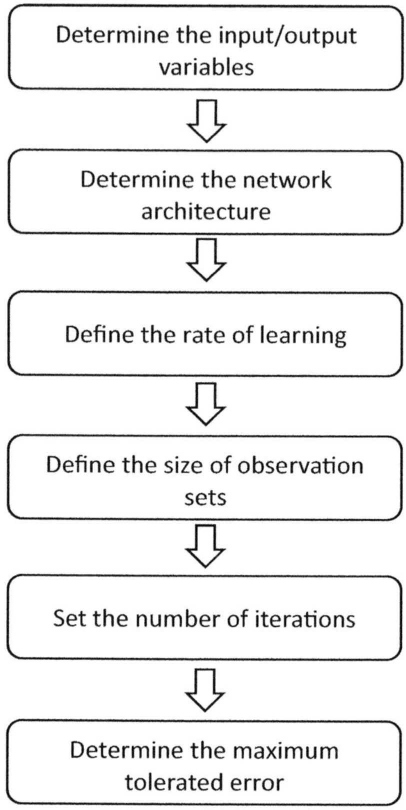

Fig. 6. Learning rules diagram

While Cadet et al. Cadet et al. (2014) have given indicators that can be used to evaluate diagnostic performances by providing precise equations to measure the degree of cell humidity. For the purpose of modeling the power supply for a (PEMFC) embedded system.

A neural network model has been created by Jeme et al. Jemeı et al. (2003) that will allow for the control of energy transfer in fuel cell vehicles.

The PEMFC dynamic model was created by Meng Shao et al. Shao et al. (2014) and is constructed and simulated using MATLAB. In order to increase the stability and reliability of PEMFC systems, the ANN ensemble for fault diagnosis (i.e., Fault in the stack cooling system; Increasing of fuel crossover; Fault in air delivery system; Fault in hydrogen delivery diagnosis) is constructed in their work.

The authors in Mammar et al. (2019) researched a method to estimate water activity in the PEMFC using the EIS and modified a neuro-fuzzy inference system (ANFIS).

In order to construct an ideal impedance model of the (PEMFC) that complies with the mass transfer theory and takes into account the physical parameters of the (EIS) model, Slimane Laribia et al. Laribi et al. (2019) have defined and put into practice a method using an artificial neural network. The authors have identified two PEM fuel cell failure modes (flooding and drying) using their ANN model. for additional studies on PEMFC diagnostics using the EIS approach.

The reader is advised to consult Lu et al. (2019); Wahdame et al. (2008); Halvorsen et al. (2020); Kahia et al. (2023); Chevalier et al. (2013); Kahia et al. (2023).

6 Comparative Study on the Prediction of PEMFC System Degradation

Different diagnostic methods have their advantages and drawbacks. The model-driven method is more robust as system characteristics change, while the data-driven approach does not require any modeling and is simple to deploy online. From the known literature so far, the data-driven methods are mainly based on the Machine Learning framework and have been investigated for the state of health of the fuel cell, many types of Machine learning approaches have been used in the predictions of the PEMFC system. The following table summarizes the characteristics of different methods (Table 1).

Table 1. Comparison of relevant and recent research on PEMFC diagnosis

Approaches	Idea	Advantages	Drawbacks	Ref
Dynamic neural network	✓ Commencing with the addition of a delay line to the (NNT)'s input to create a dynamic (NNT) control model	✓ Track the changes made to the system	✓ High computational complexity	Hatti and Tioursi (2009)
Model employing (NNT)	✓ Distinction between the normal functioning designed by a NNT and the real functioning	✓ The measurement is carried out as close as possible to the fault	✓ Difficult to decide the weights	Steiner et al. (2011)
Pattern recognition model	✓ The identification of (FCOV) based on Hamming NNT application	✓ Based on polarization curves ✓ The choose of (ΔRd) Value determine the condition of the cell	✓ Interrupt the working of the fuel cell.	Kim et al. (2012)
Modeling the power supply for a (PEMFC) embedded system	✓ Evaluate diagnostic performances by providing precise equations to measure the degree of cell humidity	✓ Used for the control loop	✓ Disturbed by the load	Cadet et al. (2014)
Neural network model	✓ Implement a software model and then integrate it into a complete vehicle powertrain in order to carry out the development and simulation of control laws	✓ Allow for the control of energy transfer in fuel cell	✓ Difficult to decide the weights	Jemei et al. (2003)
Neuro fuzzy inference system (ANFIS)	✓ Estimate water activity in the PEMFC using the EIS	✓ Accurate prediction	✓ Interrupt the working of the fuel cell long period of time	Mammar et al. (2019)
Artificial neural network	✓ Identify the (SoH) takes into account the physical parameters of the (EIS) model	✓ Construct an ideal impedance model of the (PEMFC)	✓ For EIS measurements, more equipment is needed	Laribi et al. (2019)

7 Conclusions

Artificial intelligence and machine learning are attracting increasing interest in materials development and power systems control monitoring. Machine learning informed by AI and mathematics can facilitate the development of basic knowledge, especially in the field of PEMFC technology. In this work, we have provided an overview of fuel cells' most advanced fault diagnosis and characterization methodologies. Then we described the main proprieties of learning and we have presented several related works that rely on machine learning methods for the diagnosis of PEM fuel cells.

References

Baschuk, J., Li, X.: Carbon monoxide poisoning of proton exchange membrane fuel cells. Int. J. Energy Res. **25**, 695–713 (2001)

Cadet, C., Jemei, S., Druart, F., Hissel, D.: Diagnostic tools for PEMFCs: from conception to implementation. Int. J. Hydrogen Energy **39**, 10613–10626 (2014)

Cheng, X., et al.: Hydrogen crossover in high-temperature PEM fuel cells. J. Power Sources **167**, 25–31 (2007)

Chevalier, S., Trichet, D., Auvity, B., Olivier, J., Josset, C., Machmoum, M.: Multiphysics DC and AC models of a PEMFC for the detection of degraded cell parameters. Int. J. Hydrogen Energy **38**, 11609–11618 (2013)

Dubas, F.: Conception d'un moteur rapide a aimants permanents pour l'entranement de compresseurs de piles a combustible. Ph.D. thesis. Universite de Franche-Comte (2006)

El-Kharouf, A., Mason, T.J., Brett, D.J., Pollet, B.G.: Ex-situ characterisation of gas diffusion layers for proton exchange membrane fuel cells. J. Power Sources **218**, 393–404 (2012)

Fennie, C., Reisner, D., Barbetta, J., Singh, P.: Fuzzy logic-based state-of-health determination of PEM fuel cells. In: Proceeding of Electric Vehicle Symposium and Exhibition (2001)

Fontes, G.: Modelisation et caracterisation de la pile PEM pour l'etude des interactions avec les convertisseurs statiques. Ph.D. thesis (2005)

Forrai, A., Funato, H., Yanagita, Y., Kato, Y.: Fuel-cell parameter estimation and diagnostics. IEEE Trans. Energy Convers. **20**, 668–675 (2005)

Ge, S., Yi, B., Ming, P.: Experimental determination of electro-osmotic drag coefficient in nafion membrane for fuel cells. J. Electrochem. Soc. **153**, A1443 (2006)

Halvorsen, I.J., Pivac, I., Bezmalinović, D., Barbir, F., Zenith, F.: Electrochemical low-frequency impedance spectroscopy algorithm for diagnostics of PEM fuel cell degradation. Int. J. Hydrogen Energy **45**, 1325–1334 (2020)

Hatti, M., Tioursi, M.: Dynamic neural network controller model of PEM fuel cell system. Int. J. Hydrogen Energy **34**, 5015–5021 (2009)

Hernandez, A.: Diagnostic d une pile a combustible de type PEFC. PhD thesis, Universite de Technologie de Belfort Montbelliard (2006)

Hernandez, A., Hissel, D., Outbib, R.: Fuel cell fault diagnosis: a stochastic approach. In: 2006 IEEE International Symposium on Industrial Electronics, pp. 1984–1989. IEEE (2006)

Hissel, D., Péra, M., Kauffmann, J.: Diagnosis of automotive fuel cell power generators. J. Power Sources **128**, 239–246 (2004)

Hou, M., Shao, Z., Yi, B.: Technological approaches to increasing specific power of vehicular fuel cell stacks. Strateg. Study Chin. Acad. Eng. **21**, 84–91 (2019)

Jemeı, Ä., Hissel, D., Péra, M.C., Kauffmann, J.M.: On-board fuel cell power supply modeling on the basis of neural network methodology. J. Power Sources **124**, 479–486 (2003)

Kahia, H., Aicha, S., Herbadji, D., Herbadji, A., Bedda, S.: Neural network based diagnostic of PEM fuel cell. J. New Mater. Electrochem. Syst. **23**, 225–234 (2020)

Kahia, H., Aicha, S., Herbadji, D., Herbadji, A., Belmeguenai, A.: An improved approach towards PEM fuel cell water management using artificial neural network (2023)

Kahia, H., Saadi, A., Herbadji, A., Herbadji, D., Ramadhan, H.M.: Accurate estimation of PEMFC state of health using modified hybrid artificial neural network models. J. New Mater. Electrochem. Syst. **26** (2023)

Kim, J., Lee, I., Tak, Y., Cho, B.: State-of-health diagnosis based on hamming neural network using output voltage pattern recognition for a PEM fuel cell. Int. J. Hydrogen Energy **37**, 4280–4289 (2012)

Korner, A., Tam, C., Bennett, S., Gagne, J.: Technology roadmap-hydrogen and fuel cells. International Energy Agency (IEA), Paris, France (2015)

Laribi, S., Mammar, K., Sahli, Y., Koussa, K.: Analysis and diagnosis of PEM fuel cell failure modes (flooding & drying) across the physical parameters of electrochemical impedance model: using neural networks method. Sustainable Energy Technol. Assess. **34**, 35–42 (2019)

Latham, R.A.: Algorithm development for electrochemical impedance spectroscopy diagnostics in PEM fuel cells (2004)

Li, H., et al.: Polymer electrolyte membrane fuel cell contamination: testing and diagnosis of toluene-induced cathode degradation. J. Power Sources **185**, 272–279 (2008)

Liang, D., Dou, M., Hou, M., Shen, Q., Shao, Z., Yi, B.: Behavior of a unit proton exchange membrane fuel cell in a stack under fuel starvation. J. Power Sources **196**, 5595–5598 (2011)

Lu, H., Chen, J., Yan, C., Liu, H.: On-line fault diagnosis for proton exchange membrane fuel cells based on a fast electrochemical impedance spectroscopy measurement. J. Power Sources **430**, 233–243 (2019)

Mammar, K., Saadaoui, F., Laribi, S.: Design of a PEM fuel cell model for flooding and drying diagnosis using fuzzy logic clustering. Renewable Energy Focus **30**, 123–130 (2019)

Mérida, W., Harrington, D., Canut, J., McLean, G.: Characterisation of proton exchange membrane fuel cell (PEMFC) failures via electrochemical impedance spectroscopy. J. Power Sources **161**, 264–274 (2006)

Mérida Donis, W.R.: Diagnosis of PEMFC stack failures via electrochemical impedance spectroscopy. Ph.D. thesis (2002)

Nakajima, H., Konomi, T., Kitahara, T., Tachibana, H.: Electrochemical impedance parameters for the diagnosis of a polymer electrolyte fuel cell poisoned by carbon monoxide in reformed hydrogen fuel. J. Fuel Cell Sci. Technol. **5** (2008)

Nitsche, C., Schroedl, S., Weiss, W.: Onboard diagnostics concept for fuel cell vehicles using adaptive modelling. In: IEEE Intelligent Vehicles Symposium, pp. 127–132. IEEE (2004)

O'Hayre, R., Fabian, T., Litster, S., Prinz, F.B., Santiago, J.G.: Combined heat and mass transfer model of a passive air breathing fuel cell cathode. ECS Trans. **3**, 1125 (2006)

Park, Y.H., Caton, J.A.: An experimental investigation of electro-osmotic drag coefficients in a polymer electrolyte membrane fuel cell. Int. J. Hydrogen Energy **33**, 7513–7520 (2008)

Qi, Z.: Proton Exchange Membrane Fuel Cells. CRC Press (2013)

Rubio, M., Urquia, A., Dormido, S.: Diagnosis of PEM fuel cells through current interruption. J. Power Sources **171**, 670–677 (2007)

Saber, M., Rharras, A., Saadane, R., Aroussi, H.K., Wahbi, M.: Artificial neural networks, support vector machine and energy detection for spectrum sensing based on real signals. Int. J. Commun. Netw. Inf. Secur. **11**, 52–60 (2019)

Shao, M., Zhu, X.J., Cao, H.F., Shen, H.F.: An artificial neural network ensemble method for fault diagnosis of proton exchange membrane fuel cell system. Energy **67**, 268–275 (2014)

Sharaf, O.Z., Orhan, M.F.: An overview of fuel cell technology: fundamentals and applications. Renew. Sustain. Energy Rev. **32**, 810–853 (2014)

Spiegel, C.: Designing and building fuel cells, vol. 87. Mcgraw-Hill, New York (2007)

Steiner, N.Y., Hissel, D., Moçotéguy, P., Candusso, D.: Diagnosis of polymer electrolyte fuel cells failure modes (flooding & drying out) by neural networks modeling. Int. J. Hydrogen Energy **36**, 3067–3075 (2011)

Sugiura, K., Yamamoto, M., Yoshitani, Y., Tanimoto, K., Daigo, A., Murakami, T.: Performance diagnostics of PEFC by current-pulse method. J. Power Sources **157**, 695–702 (2006)

Wahdame, B., et al.: Analysis of a PEMFC durability test under low humidity conditions and stack behaviour modelling using experimental design techniques. J. Power Sources **182**, 429–440 (2008)

Wang, X., Kumar, R., Myers, D.J.: Effect of voltage on platinum dissolution: relevance to polymer electrolyte fuel cells. Electrochem. Solid-State Lett. **9**, A225 (2006)

Yasuda, K., Taniguchi, A., Akita, T., Ioroi, T., Siroma, Z.: Platinum dissolution and deposition in the polymer electrolyte membrane of a PEM fuel cell as studied by potential cycling. Phys. Chem. Chem. Phys. **8**, 746–752 (2006)

Zhang, H., Li, J., Tang, H., Lin, Y., Pan, M.: Hydrogen crossover through perfluorosulfonic acid membranes with variable side chains and its influence in fuel cell lifetime. Int. J. Hydrogen Energy **39**, 15989–15995 (2014)

Zhang, J., Sasaki, K., Sutter, E., Adzic, R.: Stabilization of platinum oxygen-reduction electrocatalysts using gold clusters. Science **315**, 220–222 (2007)

Zhang, J., Tang, Y., Song, C., Zhang, J., Wang, H.: PEM fuel cell open circuit voltage (OCV) in the temperature range of 23 c to 120 c. J. Power Sources **163**, 532–537 (2006)

Zhang, J., Wu, J., Zhang, H., Zhang, J.: PEM fuel cell testing and diagnosis. Newnes (2013)

GPS Spoofing Attack Against UAVs: A Timeseries Dataset Case Study

Mouzai Mustapha(✉) and Riahla Mohamed Amine

M'hamed Bougara University, 35 000 Boumerdes, Algeria
{m.mouzai,ma.riahla}@univ-boumerdes.dz

Abstract. Over the past few years, the world has witnessed a notable surge in the adoption of Unmanned Aerial Vehicles in civil and military applications, including border surveillance, search and rescue, agriculture and delivery. In contrast, this potential growth has been accompanied by the lack of necessary security mechanisms that respond to the threats and vulnerabilities posed by malicious actors. Therefore, in this study we investigate one of the stealthiest attacks that afflict the navigation system of UAVs named GPS Spoofing attack. We overview the different detection techniques existing in literature, and highlight machine learning based approaches dealing with time series data.

Keywords: GPS Spoofing attack · unmanned aerial vehicles · Machine learning

1 Introduction

Unmanned Aerial Vehicles (UAVs) commonly known as drones, consist of aircraft remotely piloted either by a ground control station, or through an autopilot system following a predefined map. Thereby rendering the presence of humans onboard unnecessary.

In principle, UAVs rely on global navigation system (GPS) due to its effectiveness and worldwide coverage to determine their spatial position coordinates [1]. They measure the travelled distance of GPS signals received simultaneously from multiple satellites, by applying multilateration to obtain their precise location.

However, GPS signals are unencrypted [2] and can be easily captured, altered or retransmitted using cheap Software-defined radio tools. Malicious attackers are capable of broadcasting signals with falsified position data, affecting the UAVs navigation system to deviate from their intended path, for the purpose of hijacking or inducing them to crash.

This open field encouraged many researchers to deeply investigate vulnerabilities and security concerns related to GPS Spoofing attacks that threatens UAVs navigation system. A detailed systematic review conducted by [3] has extensively explored this threat, where authors applied multiple inclusion and exclusion criteria that discussed this topic to extract 70 pertinent papers out of 428. Subsequently, they classified them into four categories namely GPS Spoofing techniques, impact and vulnerability, GPS Spoofing as defense and finally GPS Spoofing defense mechanisms. As our interest

F. Hacène et al. (Eds.): MLN 2024, LNCS 15540, pp. 195–208, 2026.
https://doi.org/10.1007/978-3-032-00552-6_14

focuses on defense mechanisms this review detailed various proposed techniques to detect and mitigate GPS Spoofing attack in the context of UAVs.

Therefore, some works proposed to install additional onboard sensors to correct position coordinates. Authors of [4] proposed an algorithm based on Optical Flow sensor, they compared the distance travelled by the UAV with the distance between pixels of images captured using an optical flow sensor, subsequently a threshold was applied to determine the presence of GPS spoofing attack. Also, [5] employed Visual Odometry sensor to detect GPS spoofing attack, they measured the Euclidean Distance to compare the difference between various metrics of the trajectories obtained from the Odometry Cameras and the GPS. in [6], the authors designed a lightweight GPS Spoofing detection scheme in small and medium UAV formations, it consists of transmitting fused GPS receivers and Inertial Measurement Unit data of each UAV to a Ground Control station (GCS), in addition, positional information are exchanged between UAVs that collaborate with each other to accomplish a task, once the GCS notices any mismatch of the received position data, it indicates an occurring GPS spoofing attack.

Other techniques showed their effectiveness against this threat leveraging signal processing. Authors of [7] used Pseudo Range value to explore the integrity of the signals received from different satellites to eliminate the faulty ones. Meanwhile in [8], the authors proposed a 5G-assisted live detection approach against GPS spoofing, where they relied on the up-link Received Signal Strength (RSS) to verify whether a UAV position is located within a trustworthy area, otherwise it is considered as spoofed. In [9], authors investigated the detection of colluding GPS spoofing attack in UAV swarms inspired by burglary scene, they applied thresholds on Signal to Noise Ratio (C/N0) and Absolute Power to detect the attack, then they explored the Angle of Arrival (AoA) along with Pseudorandom Noise (PRN) code to locate the attacker.

Game theoretic security mechanisms have also made notable contribution in this topic, where [10] proposed a Signaling Game Framework that continuously explores the strategic interactions between spoofers and GPS receivers, until attaining a Perfect Bayesian Equilibrium (PBE). In [11], the authors proposed a defense mechanism to detect and mitigate GPS spoofing attack in UAV swarms, they formulated the exchanges between the attacker and their defending system as a Dynamic Stackelberg game, in case of GPS spoofing attack a UAV determines its position according to neighboring UAVs instead of the potential spoofed signal.

Despite the efficacity and advantages of the techniques discussed previously in countering GPS spoofing attack in the context of UAVs. They may be ineffective against novel sophisticated attacks. As artificial intelligence potentially emerge, attackers may leverage machine learning algorithms to analyze and extract patterns from authentic GPS signals, enabling them to replicate genuine localization data with falsified positions. Therefore, it becomes necessary to investigate simultaneously on sophisticated security mechanisms employing ML.

Therefore, [12] proposed an artificial neural network (ANN) to detect the legitimacy of a GPS signal, they conducted many missions using a UAV where they collected GPS signals from different locations, the features extracted included Satellite Number, Carrier Phase, Pseudo Range, Doppler Shift, and Signal to Noise Ratio. The dataset constructed comprises 2000 samples of a balanced Output distribution, where 50% of signals are legitimate and 50% are fake. Subsequently they implemented a 10-foldcrosss validation technique to evaluate their model, as a result they obtained an accuracy of 98.3%, with 99.2% probability of detection, 0.8% probability of misdetection and 2.6% probability of false alarm. However, the authors did not provide any features selection or data scaling techniques.

Researches as [13] proposed a supervised machine learning algorithm to detect the current attack, two real-time scenarios were performed to extract 13 features from the received GPS signals. Then they simulated three defined GPS spoofing attacks cases namely simple, intermediate and sophisticated attacks using MATLAB. They also applied spearman correlation technique to eliminate two irrelevant features namely RX and TCD which were highly correlated with TOW and DO features respectively, in addition this work also applied differencing to handle non-stationarity of TOW feature. The dataset obtained contains 1055 samples, where 55% of data are legitimate and 45% are spoofed samples. In further, the authors divided the dataset into 70% as training set and 30% as test set. Then, they evaluated multiple tree-based ML algorithms namely Random Forest (RF), Gradient Boost, Extreme Gradient Boost (XGBoost), and Light-GBM. The results obtained showed that XGBoost got the best results with 95.52% of accuracy, 95.38% probability of detection, 4.62% Probability of misdetection and 4.30% Probability of false alarm, additionally to other evaluation metrics such as processing time and memory size.

In [14], authors conducted the same experience as [13], they compared the performance of five instance-based ML models, namely K-Nearest Neighbors (KNN), Radius Neighbor, Support Vector Machine (SVM), C-SVM, and Nu-SVM for the same purpose. In this work, authors applied Min-Max Normalization to scale the dataset values, and differencing to handle stationarity issue of two features namely TOW and CP. Additionally they customized the learning models according to the dataset using the Grid search technique for hyperparameter optimization. Furthermore, in order to evaluate their models, 10-fold-cross-validation was applied to train 70% and test 30% of data. As a result, Nu-SVM outperformed the other models with an accuracy of 92.78%, true positive rate of 91.26%, false negative rate of 8.73% and false positive rate of 6.02%.

Authors of [15] presented a performance comparison of ensemble-learning techniques namely stacking bagging and boosting to detect GPS spoofing attack. For stacking technique they used KNN, Nearest Neighbor (NB), Decision Tree (DT), Random Forest (RF), and Logistic Regression. In bagging method, they applied DT classification. And finally, Gradient Tree Boost as boosting technique. In this work, same preprocessing techniques were conducted as [14], instead they applied Pearson Correlation as feature selection technique. They also performed 10-fold-cross-validation to train 80% and test 20% of data. The results showed that stacking based approach has the best performances with 95.43% accuracy, 99.56% probability of detection, 0.36% probability of midsection and 0.03% as probability of false alarm.

In [16], authors investigated the impact of hyperparameters, regularization parameters, correlated features, dataset sizes, and imbalanced datasets on the performance of six machine learning models namely SVM, ANN, RF, GNB, CART and LR, in detecting GPS spoofing attacks. Therefore, Pearson correlation with Chi-square feature importance were applied as features selection techniques. They concluded that inappropriate setting of the previous parameters may affect the performance of ML models. However, they didn't discuss non-stationarity issue.

In [17], the authors explored the capabilities and performances of supervised deep learning models against GPS spoofing attacks. Three DL models were implemented in this study, namely: Deep Neural Network (DNN) based on artificial neural networks, U-Neural Network (U-net) based on convolutional neural networks (CNN), and finally Long Short-Term Memory (LSTM) based on recurrent neural networks (RNN). They conducted their experiments on a balanced dataset of 7,000 authentic and 7,000 fake GPS signals and applied the Min-Max technique to normalize the data. The authors of this study highlighted how cross-validation may affect the performance of DL models and lead to overfitting. Therefore, they proposed a resampling framework to train, test, and validate their models. However, they did not provide any description of the approach. The evaluation metrics indicate that the CNN U-net model outperforms the other DL models, with 98.80% accuracy, 98.85% probability of detection, 1.15% probability of misdetection, and 1.8% probability of false alarm.

Authors of [18], presented a stacked ensemble-learning approach that leverages performance of machine learning and deep learning algorithms in detecting GPS spoofing attack. They implemented z-score normalization to scale the values of the 13 features of the dataset. Then they applied 5-fold-cross-validation to evaluate their model. The results demonstrated that stacking Machine Learning SVM model with Deep Learning CNN model attained the best performances, with an average accuracy of 99.72%, precision value of 99.65%, Recall value of 99.77% and F1-score value of 99.72% (Tables 1 and 2).

In this study, we will focus on Machine Learning based techniques. We will explore the different steps conducted to build the GPS spoofing dataset on UAVs, starting from gathering GPS features to GPS attacks simulations. Additionally, we will discuss the essential preprocessing aspects outlined in the literature, including feature selection, data scaling, model training and validation, within the context of timeseries datasets. We will also highlight the most significant open gaps existing in literature, that requires examination by researchers in the future.

The reminder of this paper is divided as follow: Sect. 2 presents analysis of techniques implemented in literature on the current dataset, Sect. 3 discusses the open gaps and leaks in previous works, Sect. 4 provides this work conclusion and mention our future direction.

Table 1. Preprocessing techniques of ML based approaches

Paper	Year	N° of used Features	Non Stationary features	Feature selection & scaling	Balancing	Validation	dataset size
[12]	2019	5	/	/	50:50	10-fold Cross Validation	2000
[13]	2021	11	TOW	- Spearnman correlation	55:45	70% train 30% test	1055
[14]	2022	11	TOW, CP	- Spearnman correlation - Min-Max normalization	55:45	70% train 30% test 10-fold cross Validation	1055
[15]	2022	11	TOW, CP	- Pearson correlation - Min-max normalization	55:45	80% train 20% test 10-fold cross Validation	1055
[16]	2022	/	/	- Pearson correlation - chi square feature Importance	50:50	80% train 20% test 10-fold cross validation	14 000
[17]	2023	11	/	- Min-max normalization	50:50	10-time resampling framework 60% train 20% test 20%validation	14 000
[18]	2024	13	/	z-score normalization	55:45	5-fold cross Validation	/

2 Machine Leaning Based Techniques Analysis

2.1 Dataset Collection

The dataset used in this work [19] contains features of real GPS signals collected from different stages of the GPS signal correlation from acquisition, tracking to the observable block. The data was collected at multiples speeds, positions, and altitudes with a customized GPS receiver of eight-parallel channels, built using a universal software radio peripheral unit (USRP), a front-end active GPS antenna, and an open-source GNSS-SDR software running on an I5-4300U laptop with 8G RAM with Ubuntu Version 16.04.7.

Table 2. Machine learning models and their results

Paper	Model	Evaluation metrics
[12]	ANN	- Accuracy: 98.3% - Pd: 99.2% - Pmd: 0.8% - Pfa: 2.6%
[13]	XGBoost	- Accuracy: 95.52% - Pd: 95.38% - Pmd: 4.62% -Pfa: 4.30% - Processing time (Training): 395.98 (ms) - Processing time (detection): 2.00 (ms) - Memory size (Training): 2.62 (MiB) - Memory size (detection): 0.01 (MiB)
[14]	Nu-SVM	- Accuracy: 92.78% - TP: 91.26% - FN: 8.73% - FP: 6.02% - Processing time (Training): 164.86 (ms) - Processing time (detection): 0.12 (ms) - Memory size (Training): 0,18 (MiB) - Memory size (detection): 0,508 (MiB)
[15]	Stacking based ensemble learning (KNN, Naïve bayes, decision tree, random forest, and logistic regression)	- Accuracy: 95.43% - Pd: 99.56% - Pmd: 0.36% - Pfa: 0.03% - Model size memory: 191.3(Mb) - Processing time: 13.06 (s) - Average prediction time per sample: 0.24 (s)
[16]	SVM, ANN, RF, GNB CART LR	- probability of detection, probability of false alarm, probability of misdetection, and accuracy with different hyperparameters, dataset size, features, and imbalanced datasets
[17]	CNN U-Net	- Accuracy: 98.80%, - Pd: 98.85%, - Pmd: 1.15%, - Pfa: 1.8%, - Training time per Sample: 0.22 s, - Prediction time:0.2 s, - Memory size: 199.87 MiB.
[18]	Stacked ensemble SVM-CNN	- Accuracy: 99.72% - Precision: 99.65% - Recall: 99.77% - F1-score: 99.72

In further, the authors gathered the features data on two steps, the first step consists of performing a flying mission from a speed of 0 Mph to 60 Mph with an average speed of 45 Mph, while the second step consist of collecting the data from three different static locations.

The dataset is available in two formats: a 3-dimentional version representing the distribution of the 13 collected features on the eight-parallel channels of the GPS receiver, and a simple 2-dimentional version converted from the 3D version according to the reception time of each channel.

2.2 Features Extraction

Thirteen features have been extracted from the GPS receiver blocks at different stages namely pre-correlation, during correlation and post-correlation, briefly described below (Table 3):

Table 3. Dataset features description

Abbreviation	Name	Stage	Description
$C/N_{\circ}$	Carrier to Noise Ratio	Pre-correlation	Ratio of the signal power to the noise in dB.
TCD	Tracking Carrier Doppler	During-correlation	Doppler shift estimated at the tracking loops in Hz.
PQP	Prompt Quadrature component	During-correlation	quadrature component of the prompt correlator (PC).
PIP	Prompt In-Phase Component	During-correlation	In phase component of the prompt correlator (PC).
PC	Prompt Correlator	During-correlation	Local correlator compares the received satellite signal with the replica signal locally generated.
LC	Late Correlator	During-correlation	1/2 chip spacing after the prompt correlator.
EC	Early Correlator	During-correlation	1/2 chip spacing before the prompt correlator.
CP	Carrier Phase Cycles	Post-correlation	beat frequency difference between the received carrier and a receiver generated carrier replica.
TOW	Time of the week	Post-correlation	the number of seconds elapsed given by the satellite atomic clock since the start of each week.
RX	Receiver Time	Post-correlation	the receiver time given in seconds after the start of time of the week (TOW).

(*continued*)

Table 3. (*continued*)

Abbreviation	Name	Stage	Description
PD	Pseudo-range	Post-correlation	Distance measured between receiver and satellite in meter.
DO	Carrier Doppler	Post-correlation	frequency drift between the sent frequency and received frequency due to doppler effect.
PRN	Pseudorandom Noise	Post-correlation	A unique identifier of each satellite.

2.3 Attacks Simulation

Since conducting real GPS attacks on autonomous aerial vehicles in outdoor environments may be illegal due to the potential risk of interfering with air traffic control systems or other aircraft operating in the vicinity. Launching such an attack indoors may also be impractical, as indoor environments lack key factors such as wind and humidity. As a result, the authors used MATLAB to simulate three GPS spoofing attacks, as defined in [20], each with varying levels of sophistication.

Simplistic Attack: All standalone civilian GPS receivers are inherently susceptible to GPS spoofing, where false signals are transmitted to deceive the target device. In its simplest form, an attacker could generate RF signals by using a GPS simulator, coupled with an amplifier and antenna. Fortunately, this attack has several drawbacks, the cost of a GPS simulator is very high, in addition it is challenging to position and hide it near the target due to its heavy weight. Moreover, launching such an attack does not synchronize with the actual GPS signal, once the attack is initiated, it causes signal tracking failure, known as "loss of lock," for the target. As a result, the target is forced to perform a process called signal reacquisition, which raises suspicions of a spoofing attack.

Intermediate Attack: This type of attack addresses the limitations of simplistic attack. An attacker can easily assemble inexpensive, off-the-shelf components to create a software-defined GPS receiver, as the structure of civilian GPS signals and software-defined radio (SDR) documentation are publicly available. Therefore, Attackers can build portable receiver-spoofers that can be discreetly placed near the target. The receiver-spoofer estimates the position, velocity, and time of the target and generates counterfeit signals that orchestrates authentic signals. Subsequently, attackers gradually increase the spoofed signal power until it gets control the delay-lock loop tracking points over the original GPS Signal. Such an attack launched from a single transmitting antenna could be detected using Angle-of-Arrival discrimination.

Sophisticated Attack: GPS spoofing defense techniques that rely on Angle-of-Arrival discrimination are often ineffective against such attack. As intermediate attacks transmit GPS signals from a single antenna, sophisticated attacks involve multiple receivers-spoofers working in collaboration to transmit spoofed signals from different angles, emulating the GPS constellation, therefore traditional detection mechanisms struggle to detect such type of attack. The most effective technique to remain against this threat is cryptographic authentication.

2.4 Data Preprocessing

Feature Selection: In this phase, it is necessary to remove the correlated and irrelevant features from our dataset. Ensemble feature selection techniques have shown effectiveness in this stage of data preprocessing, and they are classified into two classes: homogeneous and heterogeneous. Authors of [21] focused on heterogeneous ensemble feature selection using two traditional techniques, namely Spearman's Correlation and Information Gain.

$$S(f_i,\ f_j) = 1-6\sum_{1} \frac{(f_i - f_j)^2}{N\left(N^2-1\right)} \tag{1}$$

where S is the Spearman rank correlation coefficient, N represents the length of the vector, f represents the rank of feature i.

In this dataset, the pairs of features (DO, TCD) has a correlation coefficient of 94%, and (TOW, RX) has a correlation coefficient of 95%.

Information gain technique is used to measure the importance and the gain of each feature in term of the target variable of the dataset, therefore any feature that obtains a gain less than 10% will be discarded. In our case of study, DO and TCD obtained a gain of 16% and 14% respectively, while TOW and RX obtained a gain of 14% and 8.4% respectively. Consequently, TCD and RX features are discarded from the dataset.

Meanwhile, other works [16] used different techniques to measure correlation between feature alongside with their importance namely Pearson's correlation technique and Chi square feature importance

Data Scaling: Features scaling is an important step in data preprocessing since normal ranges differ from each feature, it would affect Machine Learning models performances. In [14], authors applied Min-Max normalization for features scaling, this approach consist of subtracting the minimum value in the feature and divides it by its range.

$$X = \frac{x - \mathrm{Min}(x)}{\mathrm{Max}(x) - \mathrm{Min}(x)} \tag{2}$$

where x is the initial value, Min(x) and Max(x) are the minimum and maximum values of the feature vector.

Meanwhile in other works [16, 21], they applied power transformer scalar using the Yeo–Johnson transformation technique, the latter transforms the data to fit a Gaussian distribution and handles zero, positive, and negative data values. Besides, a simple standardization technique that re-scale the values to a mean of 0 and a standard deviation of 1 known as z-score normalization [18].

$$z = \frac{(x - \mu)}{\sigma} \tag{3}$$

where z denotes the data point's Z-score, x denotes its value, μ is the dataset's mean, and σ denotes its standard deviation.

Stationarity: Since the dataset collection process involves conducting multiple missions utilizing a UAV, and gathering GPS signals over time. Therefore, we consider it as time series dataset, which indicate that the data samples lack static relationships including mean, median, and variance. In this case traditional ML algorithms are unable to learn and extract patterns from these situations referred to non-stationarity. To overcome such issue, we need to implement a process that consist of calculating the consecutive differences between samples commonly known as Differencing. Therefore, [14] indicated that TOW and CP features are not stationary, thus it is necessary to apply differencing on these two features.

$$\Delta x_i = x_{i+1} - x_i \tag{4}$$

where Δx_i is the first order differentiation, x_i is a data sample in the non-stationary data.

2.5 Models Validation and Evaluation

Various Machine Learning and Deep Learning algorithms have been trained and tested on this dataset, where the target feature was adjusted to be exploited as binary classification problem, therefore 0 represent legitimate signals and 1 represent spoofed signals. The dataset, provided in a 2D format by the authors, was imbalanced due to the fact that generating fake signals is more complex than collecting authentic GPS signals. Consequently, existing literature often only extracts a small portion of the dataset to train models in a balanced environment. In addition, timeseries datasets require a specific approach for train-test split process, since random splitting is not appropriate in such cases. The test set must always follow the training set, because models cannot be trained on future data to predict data from the past.

Meanwhile, to validate ML models, most of works applied a validation technique known as k-fold cross validation, this technique consists of splitting the data into k folds, and each time train the model on the firsts (i) th folds and test it with the (i) th + 1 fold (Fig. 1).

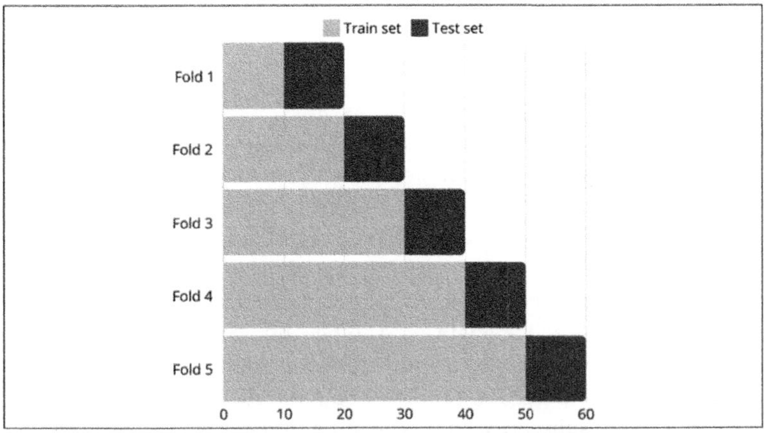

Fig. 1. K-fold cross validation in timeseries datasets, K = 5

As for evaluation metrics, most of works evaluated their models in terms of accuracy, probability of detection, probability of midsection and probability of false alarm, additionally as UAV has limited hardware capabilities including processors and batteries, other works investigated other performance metrics such as processing power, memory size and prediction time [15].

$$Accuracy = \frac{(TP + TN)}{(TP + TN + FP + FN)} \tag{5}$$

$$Pd = \frac{(TP)}{(TP + FN)} \tag{6}$$

$$Pmd = \frac{(FN)}{(TP + FN)} \tag{7}$$

$$Pfa = \frac{(FN)}{(TP + FN)} \tag{8}$$

where Accuracy is the measure of how well a model correctly predicts instances, Probability of detection is the ability of a model to correctly identify positive instances, probability of misdetection is how often a model fails to detect positive instances, and finally probability of false alarm which is the measure a model incorrectly identifies negative instances as positive.

Meanwhile, TP is the number of correct predicted malicious attacks referred to as True Positive, TN is the number of correct predicted normal attacks referred to as True Negative, FP is the number of incorrect predicted malicious attacks referred to as False Positive, and FN is the number of incorrect normal attacks referred to as False negative.

3 Open Gaps

3.1 Dataset Challenges

Despite many studies exploited this dataset showed promising results, nonetheless it is crucial to explore more enhancements. Firstly, it only contains thirteen features divided into different correlation stages, one feature in the pre-correlation, six features during-correlation and six features in the post-correlation stage, therefore we recommend researchers to include additional features, either by extracting them from GPS receivers or by using techniques such as feature engineering. More features would help to delve deeper into GPS attack analysis. Another important key point is to investigate the pre-correlation features, this would allow for early identification of an occurred GPS spoofing attack, thus the detection time will be reduced. Such an approach could also trigger suspicion in the subsequent stages helping in reducing false alarms.

The data collection process comprised two scenarios: the first scenario involved collecting data from different static locations, meanwhile the second scenario was a driving test. Consequently, the dataset was leaked to real case missions especially in different weather conditions such as windy, cloudy and rainy weathers, which affect the quality of GPS signals. This confirms the first key point, adding weather features such as wind speed, humidity, temperature and other parameters that change over time may help to conduct more reliable datasets.

Furthermore, the GPS receiver constructed by the authors comprises eight parallel channels, each channel processes incoming signals from a single satellite.

However, the reasons for selecting an eight-channel GPS receiver are not stated. it is crucial to investigate the detection of GPS spoofing attacks in UAVs equipped with various number of channels, since increasing channels number decreases the power savings [22] and vice versa. Consequently, it is essential to develop effective spoofing detection solutions that consider the choice of channel number to address the challenges of power consumption.

3.2 Time Series Data Challenges

As the features in the studied dataset collected over a time interval, it is important to specify that the dataset is considered as a time series case, which requires careful pre-processing techniques to account the temporal patterns present in the data. Few works have mentioned the existence of non-stationary features in this dataset [14, 15]. Meanwhile it is known that ML algorithms cannot extract correlation between samples in a non-stationary scenario, due to the fact that statistical properties such as mean, variance, covariance, and standard deviation change over time.

Therefore, it is necessary to apply techniques such as differentiation to address this issue. However, previous works identified stationarity using visual plotting. Hence, it is crucial to implement more verification methods, such as statistical tests like the Augmented Dickey-Fuller (ADF) test.

In addition, the 2D format of the dataset is unbalanced, so many studies extracted a smaller subset of the data to deal with this concern. However, it is important to mention that extracting data pieces from a timeseries dataset, should not be done randomly, rather, it must preserve the temporal order to maintain the integrity of the data's time dependent structure.

Most studies used k-fold-cross-validation, typically using variations of 3, 5, or 10 folds. However, [17] pointed out that k-fold cross-validation can lead to overfitting in time series datasets since the splitting process should respect the chronological order. Furthermore, the timestamps in this dataset are not constant, instead they used a 10-time resampling framework, to divide the data into 60% as a train set, 20% as test set, and 20% as validation set. Authors avoided applying k-fold cross-validation due to likely overfitting, however, they did not provide further details of the approach used.

3.3 Conducting GPS Spoofing Attacks Challenges

In this dataset, the authors simulated three GPS spoofing attacks using MATLAB, They adjusted the values of each feature based on the impact that has each attack previously defined in this paper on them. With recent advancement in ML algorithms capabilities, it is important to further explore GPS spoofing attacks that leverage the power of ML models. They could be employed to predict the position of targets without being physically near, as is the case with receiver-spoofer-devices used in intermediate and sophisticated attacks. Additionally, they could also be applied to predict the strength of the received signals, to overcome the limitations of the simplistic attack.

However, there still exist gaps in the literature in terms of realistic GPS spoofing attacks datasets in the context of UAVs, therefore more researches need to be investigated on this topic.

4 Conclusion

This paper presents a significant vulnerability that threatens Un-manned Aerial Vehicles (UAVs), known as GPS spoofing attack. This type of attack involves misleading the Global Navigation System in order to hijack or crash UAVs. We described three principal categories of this attack namely simplistic, intermediate and sophisticated attacks, alongside the necessary steps and tools needed. Furthermore, in order to address this issue, we have reviewed several countermeasures approaches available in the literature including signal processing, onboard devices, game theory and machine learning. In our study, we examined Machine Learning based techniques at different stages, beginning with the collection of GPS data, then passing to preprocessing functions that handle time series challenges, and lastly ML and DL models existing in the literature that contribute in detecting the attack. These models were evaluated using different metrics typically required in UAVs environments, such as detection accuracy, processing time and memory size. In conclusion, we identified the main gaps that requires further investigation. The dataset explored in our work lacks additional features, with more realistic GPS spoofing attacks at different external conditions. Additionally, it would be beneficial to conduct further analysis of Machine learning and Deep learning models that handle time series data more effectively. In future work, the aim is to explore more features utilizing feature engineering techniques, and to delve deeply into powerful models that process time patterns.

References

1. Isik, O.K. et al.: Integrity analysis for GPS-based navigation of UAVs in urban environment. Robotics **9**(3), 66 (2020)
2. Heng, L., et al.: GPS signal authentication from cooperative peers. IEEE Trans. Intell. Transp. Syst. **16**(4), 1794–1805 (2015)
3. Altaweel, A., Kamel, I.: GPS spoofing attacks in FANETs: a systematic literature review. IEEE Access. **11**, 55233–55280 (2023)
4. Feng, Z., et al.: Efficient drone hijacking detection using onboard motion sensors. In: Design, Automation, and Test in Europe, pp. 1418–1423. IEEE (2017)
5. Varshosaz, M., et al.: Spoofing detection of civilian UAVs using visual odometry. ISPRS Int. J. Geo-information **9**(1), 6 (2019)
6. Liang, C., et al.: Detection of global positioning system spoofing attack on unmanned aerial vehicle system. Concurr. Computat. Pract. Exp. (2020)
7. Javaid, A.Y., et al.: Analysis of global positioning system-based attacks and a novel global positioning system spoofing detection/mitigation algorithm for unmanned aerial vehicle simulation **93**(5), 427–441 (2017)
8. Dang, Y., et al.: GPS spoofing detector with adaptive trustable residence area for cellular based-UAVs. In: Global Communications Conference, pp. 1–6. IEEE (2020)
9. Bada, M., et al.: A policy-based solution for the detection of colluding GPS-Spoofing attacks in FANETs. Transport. Res. Part A-policy and Pract. **149**, 300–318 (2021)
10. Zhang, T., Zhu, Q.: Strategic Defense Against Deceptive Civilian GPS Spoofing of Unmanned Aerial Vehicles. https://link.springer.com/chapter/10.1007%2F978-3-319-68711-7_12
11. Eldosouky, A., et al.: Drones in distress: a game-theoretic countermeasure for protecting UAVs against GPS spoofing. IEEE Internet Things J. **7**(4), 2840–2854 (2020)
12. Manesh, M.R., et al.: Detection of GPS spoofing attacks on unmanned aerial systems. In: Consumer Communications and Networking Conference, pp. 1–6. IEEE (2019)
13. Aissou, G.: Tree-based supervised machine learning models for detecting GPS spoofing attacks on UAS. IEEE 12th Annual Ubiquitous Computing, Electronics & Mobile (2021)
14. Aissou, G., et al.: Instance-based supervised machine learning models for detecting GPS spoofing attacks on UAS. In: Computing and Communication Workshop and Conference, pp. 0208–0214 (2022)
15. Gasimova, A., et al.: A comparative analysis of the ensemble models for detecting GPS spoofing attacks on UAVs. In: Computing and Communication Workshop and Conference, pp. 0310–0315 (2022)
16. Khoei, T.T., et al.: Impact of dataset and model parameters on machine learning performance for the detection of GPS spoofing attacks on unmanned aerial vehicles. Applied Sciences **13**(1), 383 (2022)
17. Khoei, T.T., et al.: Supervised Deep Learning Models for Detecting GPS Spoofing Attacks on Unmanned Aerial Vehicles, 340–346 (2023)
18. Eshmawi, A., et al.: Enhanced machine learning ensemble approach for securing small unmanned aerial vehicles from gps spoofing attacks. IEEE Access (2024)
19. Aissou, G., et al.: A dataset for gps spoofing detection on autonomous vehicles. https://ieee-dataport.org/documents/datasetgps-spoofing-detection-autonomous-vehicles (2022)
20. Assessing the Spoofing Threat: Development of a Portable GPS Civilian Spoofer. In: 21st International Technical Meeting of the Satellite Division of The Institute of Navigation. Todd E. Humphreys (2008)
21. Khoei, T.T., et al.: Dynamic selection techniques for detecting GPS spoofing attacks on UAVs. Sensors **22**(2), 662 (2022)
22. Sathyamorthy, D., et al.: Evaluation of the trade-off between Global Positioning System (GPS) accuracy and power saving from reduction of number of GPS receiver channels. Applied Geomatics **8**(2), 67–75 (2016)

Author Index

A
Abdelhamid, Hocine 1
Ahmed, Sajid 149
Alhachem, Christina 32
Amine, Riahla Mohamed 195

B
Badis, Hakim 133
Belgacem, Ali 73
Belmeguenai, Aissa 180
Berbineau, Marion 133
Berri, Sara 55
Bouabdallah, Abdelmadjid 32
Boumerdassi, Selma 180
Bradai, Abbas 73

C
Chorti, Arsenia 55
Conan, Vania 1

D
Djema, Mohamed Yacine 97

F
Flauzac, Olivier 97
Fouchal, Hacene 97

H
Hachem, Nagham 116
Hamamouche, Said 14
Hammamouche, Assia 14
Herbadji, Abderrahmane 180
Herbadji, Djamel 180

J
Jia, David 1
Joly, Morgane 85

K
Kahia, Hichem 180
Kellil, Mounir 32
Khoukhi, Lyes 14

L
Leghris, Cherkaoui 170
Liu, Dingyang 133

M
Maaloul, Sassi 133
Mehaoua, Ahmed 102
Mendiboure, Leo 133
Molla, Dereje Mechal 133
Mustapha, Mouzai 195

N
Nabil, Oussama 170

R
Renault, Éric 85, 116
Rivière, Fabian 85

S
Salem, Osman 102

T
Teo, Sin G. 55
Trinh-Nguyen, Bac 55
Truong-Huu, Tram 55

X
Xiang, Jiahui 102

Y
Yoshiura, Noriaki 149

Z
Zamouche, Djamila 14

F. Hacène et al. (Eds.): MLN 2024, LNCS 15540, p. 209, 2026.
https://doi.org/10.1007/978-3-032-00552-6

GPSR Compliance

The European Union's (EU) General Product Safety Regulation (GPSR) is a set of rules that requires consumer products to be safe and our obligations to ensure this.

If you have any concerns about our products, you can contact us on ProductSafety@springernature.com

In case Publisher is established outside the EU, the EU authorized representative is:

Springer Nature Customer Service Center GmbH
Europaplatz 3
69115 Heidelberg, Germany

Batch number: 09078531

Printed by Printforce, the Netherlands